Plastics in the Environment

Edited by Alessio Gomiero

Published in London, United Kingdom

IntechOpen

Supporting open minds since 2005

Plastics in the Environment
http://dx.doi.org/10.5772/intechopen.75849
Edited by Alessio Gomiero

Contributors
Katrin Schuhen, Michael Sturm, Adrian Herbort, Alessio Gomiero, Pierluigi Strafella, Gianna Fabi, Daniela Berto, Federico Rampazzo, Claudia Gion, Seta Noventa, Malgorzata Formalewicz, Francesca Ronchi, Umberto Traldi, Giordano Giorgi, Elen Pacheco, Denis Ribeiro Dias, Maria Guimarães, Christine Nacimento, Célio Albano, Geovanio Oliveira, Mônica Andrade, Ana Sousa, Ana Silva, John A Glaser

First published in London, United Kingdom, 2019 by IntechOpen
IntechOpen is the global imprint of INTECHOPEN LIMITED, registered in England and Wales, registration number: 11086078, The Shard, 25th floor, 32 London Bridge Street
London, SE19SG – United Kingdom
Printed in Croatia

British Library Cataloguing-in-Publication Data
A catalogue record for this book is available from the British Library

Additional hard copies can be obtained from orders@intechopen.com

Plastics in the Environment
Edited by Alessio Gomiero
p. cm.
Print ISBN 978-1-83880-492-3
Online ISBN 978-1-83880-493-0
eBook (PDF) ISBN 978-1-78984-045-2

For EU product safety concerns:
IN TECH d.o.o., Prolaz Marije Krucifikse Kozulić 3, 51000 Rijeka, Croatia, info@intechopen.com or visit our website at intechopen.com.

Meet the editor

Alessio Gomiero holds a PhD in Environmental Science. As a senior researcher, he has contributed to the development of integrated chemical and biological tools to routinely assess the environmental pollution associated with different anthropogenic activities. Emphasis is placed on the characterization of chemical-induced adverse effects by cellular and subcellular endpoints on several marine and freshwater biological models.

Contents

Preface

This book is a collection of reviewed and relevant research chapters concerning developments within the plastics in the environment field of study. The book includes scholarly contributions by various authors and is edited by experts pertinent to plastic pollution. Each contribution comes as a separate chapter complete in itself but directly related to the book's topics and objectives.

The book consists of five chapters: (Chapter 1) From Macroplastic to Microplastic Litter: Occurrence, Composition, Source Identification and Interaction with Aquatic Organisms. Experiences from the Adriatic Sea, (Chapter 2) Technological Approaches for the Reduction of Microplastic Pollution in Seawater Desalination Plants and for Sea Salt Extraction, (Chapter 3) Elemental Analyzer/Isotope Ratio Mass Spectrometry (EA/IRMS) as a Tool to Characterize Plastic Polymers in a Marine Environment , (Chapter 4) Study of the Technical Feasibility of the Use of Polypropylene Residue in Composites for Automotive Industry and (Chapter 5) Biological Degradation of Polymers in the Environment.

This book will be interesting to various readers, researchers, scholars, and specialists in the field, who will find this information useful for the advancement of their research work.

IntechOpen

Chapter 1

From Macroplastic to Microplastic Litter: Occurrence, Composition, Source Identification and Interaction with Aquatic Organisms. Experiences from the Adriatic Sea

Alessio Gomiero, Pierluigi Strafella and Gianna Fabi

Abstract

Marine litter is human-created waste that has been discharged into the coastal or marine environment. "Marine debris" is defined as anthropogenic, manufactured, or processed solid material discarded, disposed of, or abandoned in the environment, including all materials discarded into the sea, on the shore, or brought indirectly to the sea by rivers, sewage, storm water, waves, or winds. A large fraction of marine debris is made up of plastic items. Plastic marine debris has become one of the most prevalent pollution related problems affecting the marine environment globally. The widespread challenge of managing marine litter is a useful illustration of the global and transboundary nature of many marine environmental problems. At a global level, plastic litter constitutes 83–87% of all marine litter. Land-based sources are estimated to be responsible for approximately 80% of marine litter. The largest portion of plastic associated with marine pollution is often linked to the contribution from terrestrial sources associated with accidental or deliberate spills as well as inefficient waste management systems in heavily anthropized coastal regions. This chapter is intended to serve as a catalyst for further discussion to explore the potential for developing a Mediterranean regional framework for addressing marine litter.

Keywords: plastic debris, Adriatic Sea, sediments, floating litter, sediments, distribution models

1. Introduction

We live in the "Plastic Age". From its creation in the early 1870, plastic material has largely contributed to the society development making everyday life easier. Plastic material offer good advantages as it can be customized with specific shapes and chemical and physical properties i.e., elasticity, hardness, lightness, transparency and durability. Due to this, the production has dramatically boosted annual plastic production from 0.5 million tons in the 40s to 550 million tons in 2018 [1].

However, plastics sturdiness presents some negative implications as the increasing rate of plastic consumption worldwide its release in the environment associated with a low degradation rate is resulting in its accumulation in coastal and marine sediments, pelagic and benthic biota from coastal to open ocean areas at each latitude from the poles to the equator. Depending on sources and formation mechanisms plastic fragments are split into "primary" and "secondary". Primary plastics are resulting from the direct input of freshly manmade emissions, adding new micronized size by-design plastic material to the environment. According to this definition, major sources primary plastics are: (A) polymers intentionally produced and used as such. In this group belong i.e., personal care consumer products, industrial or commercial products and other specialty chemicals with plastic microbeads; (B) inherent collateral products of other industrial activities or (C) plastic sourced as accidental or deliberate spillage i.e., pellets loss from plastic factories and transport. In contrast, secondary plastics are associated as secondary pollution sources where larger plastic items undergo degradation and subsequent fragmentation leads to the formation of smaller plastic pieces as they start to break down by photo-oxidative degradation followed by thermal and/or chemical degradation [2].

2. Sources, degradation processes, detection of plastic debris in marine environments

While addressing the comprehension of plastics degradation mechanisms in marine aquatic environments it is useful to divide them into plastics with a carbon-carbon backbone and plastics with heteroatoms in the main chain. Some of the most environmentally recurrent polymers like polyethylene, polypropylene, polystyrene and polyvinylchloride have a pure carbon-based backbone. On the contrary, polyethylene terephthalate and polyurethane plastics have heteroatoms in the main chain. Most packaging materials are made of plastics with a carbon-carbon backbone structure. As they are very often discarded after a short period of time, there is a high potential to observe significant loading in the environment. All these polymers are susceptible to photo-initiated oxidative degradation, which is believed to be their most important abiotic degradation pathway in aerobic outdoor environments. This degradation pathway consists of a complex sequential multi-step process where initially chemical bonds in the main polymer chain are broken down by light, by heat or by a combination of both to produce a free radical formation [3, 4]. Polymer radicals react with oxygen and form a peroxy-radical species. As a side effect, the co-occurring formation of hydroperoxides promotes a further complex pathway of radical reactions leading to significant autoxidation of the target polymer. These processes ultimately lead to chain scission, branching and creation of oxygen-containing functional groups. As the molecular weight of the polymers is reduced, the material becomes fragile and is more vulnerable to fragmentation, which makes a higher surface area reactive to further degradation. Nevertheless, anti-oxidants and stabilizers used as additives inhibit the degradation of the polymer. Thus, degradation rates depend strongly on used additives and plasticizers [4]. In most cases these are well-known toxic chemicals not covalently bonded to the polymer and therefore capable of leaching out from the plastic during the degradation process, and easily enters into the aquatic environment representing a further point of concern for eco-toxicologists. On the other hand, different degradation mechanisms cause degradation of plastics with heteroatoms in the main chain. They show an increased thermal stability compared to polymers with a simple carbon backbone. Under marine environmental conditions the degradation processes of plastics like polyethylene terephthalate (PET) or polyurethane (PU) are normally controlled by hydrolytic cleavage. Similar to carbon-carbon backbone plastic polymers, PET can undergo photo-induced autoxidation via radical reactions leading to the ultimate formation of a carboxylic acid end groups, which show a promoting effect on thermo- as well as photo-oxidative degradation. Weathering of PET in the marine environment occurs mainly by photo-induced oxidation and secondly by hydrolytic degradation processes which cause the yellowing of the polymer. For thermo-oxidative degradation the consequences are an in the content of the some end groups i.e., carboxylic acid as well as a general decrease in molecular weight of the main polymer [4]. Hydrolysis also leads to a reduction in molecular weight and an increase in carboxylic acid end groups. PET is highly resistant to environmental biodegradation because of its compact structure [4]. On the other hand, polyurethane-like compounds show carbon, oxygen and nitrogen in the main chain demonstrating enhanced susceptibility to degradation via photo-oxidation, hydrolysis and biodegradation. Plastic floating on the ocean surface is exposed to moderate temperatures, solar radiation at wavelengths of 300 nm and longer, as well as oxidizing conditions. Since temperatures are moderate, the most important factors initiating abiotic degradation are oxygen and sunlight. According to recent studies, fragmentation patterns first occur at the plastic surface, which is exposed and available for chemical or photo-chemical attack. The process is more efficient with smaller plastic fragments as they show a higher surface to volume ratio [5]. Changes in color and crazing of the surface are the initial visual effects of polymer degradation. Surface cracking makes the inside of the plastic material available for further degradation, which eventually leads to embrittlement and disintegration. Furthermore, almost all commercial plastics include additives. These co-production chemicals embedded in the polymers can also leach into the aquatic environment, which is an additional point of concern. As these substances enhance plastics' resistance to degradation, it becomes difficult to quantitatively estimate the fragmentation patterns since different plastic products can vary in their composition. On the other hand, additional factors can significantly influence degradation rates as floating plastic may develop biofilms that shield it from UV radiation. The formation of biofilm in plastic microliter collected from the marine aquatic environment has been previously documented worldwide [6–8]. Such phenomena could lead to a reduction in photo-initiated degradation. So far, there have been very few studies of degradation mechanisms for plastic polymers in the marine environment although some promising early findings have been reported by ongoing joint research initiatives (e.g., JPI-Weather Mic and JPI-PlasTox). The biofilm formation can also affect the vertical distribution of plastic fragments largely affecting their distribution in the water column or in the sedimentary environment. Most synthetic polymers are buoyant in water and substantial quantities of plastic debris that are buoyant enough to float in seawater are transported and potentially washed ashore. The polymers that are denser than seawater tend to settle near the point where they entered the environment; however, they can still be transported by underlying currents. **Table 1** resumes the theoretical densities of the most recurring polymers found in the environment. Microbial films rapidly develop on submerged plastics and change their physico-chemical properties such as surface hydrophobicity and buoyancy [9, 10]. All in all, plastic debris is a mixture of molecules and chemicals, its size ranging from some meters to a few micrometers and probably nanometers. It is derived from a broad variety of origins, such as fishing gear, nets, bottles, bags, food packaging, taps, straws, cigarette butts and cosmetic microbeads and the associated fragmentation of all of these. Plastic debris has become ubiquitous in all environmental compartments of the marine ecosystem form sediments to sea surface. Thus, the observed loadings floating in the ocean represents only a limited portion of the total input. It has been previously reported that most plastic litter ends up on the seabed with a remaining fraction distributed on beaches or floating on the seawater surface leading one to

Polymer	Abbreviation	Density (g/cm^3)	Applications
Expanded polystyrene	EPS	0.01–0.04	Bait boxes, floats, cups
Low density polyethylene	LDPE	0.89–0.93	Plastic bags, bottles, gear, cages
High density polyethylene	HDPE	0.94–0.98	Plastic bags, bottles, gear, cages
Polypropylene	PP	0.83–0.02	Rope, bottle caps,
Polypropylene terephthalate	PET	0.96–1.45	Bottles, gear
Styrene butadiene rubber	SBR	0.94	Car tyre
Polyamide	PA	1.02–1.16	Gera, fish farm nets, rope
Polystyrene	PS	1.04–1.10	Containers, packaging
Polymethyl methacrylate	PMMA	1.09–1.20	Insulation, packaging
Polyvinylchloride	PVC	1.16–1.58	Film, pipe, containers
Polycarbonate	PC	1.20–1.22	Textiles, leisure boats
Polyurethane	PU	1.20	Insulation, floats
Alkyd	ALK	1.24–2.10	Paints, packaging
Polyester	PES	1.24–2.3	Textiles,
Polytetrafluoroethylene	PTFE	2.1–2.3	Personal care products

Table 1.
Theoretical densities of the most recurring polymers found in the environment.

consider that merely quantifying floating plastic debris may lead to a significant underestimation of the actual amount of plastics in aquatic environments [11].

2.1 The interaction of plastic debris with aquatic life

Overall ecosystem health can be significantly affected by the accumulation of trash and plastics in our seas. Ingestion of and entanglement in marine debris directly impacts marine life. Laboratory studies provide a strong proof of evidence for the effects of microplastic ingestion observed in organisms collected from the natural environment. Indeed, in laboratories, under natural like conditions, microplastics have been shown to be ingested by amphipods, barnacles, lugworms and bivalves [12–14]. In the same organisms, the uptake of microplastics caused notable ultrastructural changes in the investigated tissues including histological changes as well as cell functioning impairments [15]. In field observations, the occurrence of MPs in the gastrointestinal tract and gills of pelagic and demersal fish and marine mammals has been documented [16, 17]. Past reports have shown that many marine organisms wrongly identify plastic debris for food. Ingestion of marine debris induce different deleterious effects such as pathological alteration, starvation and mechanical blockages of digestive processes. Furthermore, the interaction of plastic fragments, especially those at micrometric and nanometric scales, with organic pollutants are of importance in relation to environmental contamination and biological effects on organisms in the water column as well as in the sedimentary environment [18, 19]. Hydrophobic pollutants co-occurring in the aquatic environment may in fact adsorb onto MP debris. According to the different sizes, plastic fragments have the potential to transport contaminants more effectively through biological membranes and ultimately inside cells of aquatic organisms. The presence of organic pollutants on marine plastics has been illustrated for a wide range of chemicals in natural aquatic conditions [20, 21]. The exposure routes of organic pollutant-enriched MPs are varied, while the toxicity is largely inversely correlated to the size of the particles, as the smaller the particle the further into the organism it can penetrate releasing toxic chemicals under acidic gut conditions [22]. According to the properties of the adsorbed chemicals, several toxicity mechanisms are represented by increased oxidative stress, genotoxicity, depletion of immune competence, impairment of key cell functioning, loss in reproductive performance, disorders in energy metabolism, and changes in liver physiology [23–25].

2.2 Extracting microplastics from environmental matrices

Different methods have been developed for identifying plastics, including meso, micro and nanoplastics in water, sediments and biota as well as to a lesser extent in soil. The percentage of organic matter (OM) in general as well as some recurring specific macromolecules, such as fats and proteins may hamper the analysis, thus hiding plastic fragments in visual analyses and distort signals in Fourier transformed infrared (FT-IR) and Raman spectroscopy, two of the most frequently used methods for plastic identification [26, 27]. Hence, identifying and quantifying plastic materials in organic matter enriched samples may be a challenge. In sediments, several available protocols recommend a preliminary sorting of plastic size grounding and sieving. After sieving, the mineral phase of soils might be removed easily using density fractionation methods. Different density solutions have been used including NaCl, $ZnCl_2$, NaI and more recently $3Na_2WO_4$ $9WO_3$ H_2O to obtain dense floating solutions [28, 29]. However, it has been shown that simple density fractionations will not succeed in separating organic matter from plastic materials in sediments because most of the OM show densities between 1.0 and 1.4 g/cm^3, similar to that of several environmentally recurring plastic types like PET, PP, PE and Nylon. Sufficient removal of OM without destroying small plastic polymers is challenging because large parts of OM are refractory. At the same time, polymers show strong sensitivity to acidic or strong oxidizing treatment conditions, which induce permanent modifications (e.g. yellowing), thus hampering their classification by microscope-oriented techniques. To efficiently remove OM, multistep extraction, purification processes based on alkaline treatments possibly combined with multi-enzymatic digestion steps have been suggested for the analyses of biota water or sediments. Enzymatic digestion has been promising for the removal of organic as well as other interferents, such as chitin, agar and lipid enriched samples [27]. Strong alkali digestions have been pointed out as being effective for sediments as well as biological samples, without altering the plastic itself [30]. While on the contrary and as previously mentioned, strong acidic conditions induce partial dissolution of polycarbonate as well as partial digestion of polyethylene and polypropylene [13]. Another largely exploited strategy to remove organic matter relies on the application of concentrated hydrogen peroxide [26]. However, its use must be critically evaluated in terms of digestion conditions as treatments with incubation exceeding 48 h with temperatures exceeding 50C, which may degrade plastic polymers like polyethylene and polypropylene [31]. In this context, some authors have recently suggested an effective combined multistep method based on a sequence of enzymatic digestions followed by a short hydrogen peroxide treatment for the removal of organic matter from complex environmental matrices (e.g., wastewater samples). In summary, several promising methods have been tested for extracting, purifying and pre-concentrating plastic materials from sediments and marine biota, all of them having potential limitations. More research is needed to develop a standard protocol for isolating plastics from a range of different environmental matrices, ideally at low cost and without altering plastic properties.

2.3 Overview of the most applied detection and quantification methods

Once isolated, plastic fragments can be tracked and characterized by different analytical techniques. Some are defined as "surface oriented" methods like Raman spectroscopy, Fourier Transformed Infra-Red (FTIR), Scanning Electron Microscopy/Energy Dispersive X-Ray Spectroscopy (SEM-EDS) and environmental scanning electron microscope (ESEM) with an attached X-ray energy dispersive system (ESEM-EDS). Plastic fragments are visually sorted and analyzed coupled with microscopy. However, as discussed above, the use of strong oxidant/acidic agents applied during the extraction from sometimes complex environmental matrices (e.g., organic matter enriched marine sediments, or fat rich marine biota) may induce alteration in the plastic surface like partial dissolution, yellowing and polymer structure disruption leading to erroneous characterization of microparticles. Furthermore, some compounds of natural origin occurring in marine samples (e.g., chitin) have shown spectroscopic properties similar to those of the most recurrent plastic polymers leading to inaccurate polymer characterizations and overall abundance estimation. In addition, these microscopy-based techniques are time consuming and unable to process large numbers of samples. However, significant advances in the automatic and semi-automatic FTIR spectra recognition have been recently presented as promising time saving solutions (Jes recent paper). Alternatively, promising solutions include the Pyrolysis-gas chromatography in combination with mass spectrometry (Pyr-GC-MS) as well as the Thermogravimetric analysis coupled with mass spectrometry (TGA-MS). Pyr-GC-MS in particular can be used to assess the chemical composition of potential microplastic particles by analyzing their thermal degradation products. The polymer origin of particles is identified by comparing their characteristic combustion products with reference pyrograms of known virgin-polymer samples. Py-GC/MS had the advantage of being able to analyze the polymer type and OPA content in one run without using any solvents and with few background contaminations. Additionally, the Pyr-GC/MS method has an appropriate degree of sensitivity for analyzing plasticizers in microplastic particles with limited sample masses. However, although the pyrolysis-GC/MS approach allows for a good assignment of potential microplastics to polymer type it has the disadvantage of being a "destructive" technique as the sample is burned to obtain the pyrolytic products. Furthermore, due to limitations in the quantity of sample loaded in the pyrolysis cup only particles of a certain minimum size can be processed resulting in a lower size limitation of particles that can be analyzed. Each of these methods have their own limitations and advantages, therefore, their combined use, especially for the analysis of complex environmental samples, is a recommended strategy to reduce the effect of interferents in the analysis and obtain reliable results.

3. The Mediterranean and the Adriatic Sea

With some of the most significant amounts of solid waste generated annually per person (208–760 kg/year), the Mediterranean Sea is one of the world's areas most affected by litter [32]. The estimated amount is 62 million of macrolitter items floating on the surface of the whole basin [33]. Litter enters the seas from land-based sources, ships and other infrastructure at sea and can travel long distances before being deposited on the seabed or along the coasts. Mean densities of floating microplastics in the Mediterranean Sea of more than 100,000 items/km^2 [34] indicate the importance of this threat for the basin. In this context, the Adriatic Sea represents a hot spot for plastic litter both because of peculiarities in its oceanographic conditions as well as the high degree of anthropogenic pressure related to tourism, artisanal and industrial activities coexisting in a narrow area. The Adriatic Sea is an elongated basin, located in the central Mediterranean, between the Italian peninsula and the Balkans, with its major axis in the NW-SE direction. The northern area is very shallow, gently sloping, with an average depth of about 35 m, while the central part is on average 140 m deep, with the two Pomo depressions reaching 260 m. The northern and central parts of the basin are affected by a great number of rivers along the Italian coast, of which the Po river is the most relevant. River discharge and wind stress are the main drivers of the water circulation. West Adriatic Current (WAC), flowing SE along the western coast, and East Adriatic Current (EAC), flowing NE along the eastern coast are the main currents affecting the Adriatic circulation. There are two main cyclonic gyres, one in the northern part and the other in the south. The Bora wind (from NE) causes free sea surface to rise close to the coast enhancing the WAC and the Sirocco wind (from SE), which is the major wind affecting the Adriatic Sea, leads flood events in the shallow lagoons along the basin coast [35]. A vertical thermohaline front parallel to the coast and extending throughout the water mass, divides the coastal waters from the open sea. This retains the materials flowing from rivers and other water sources within the coastal area. A stratification characterizes the water column separating the warmer surface waters with lower salinity from deeper, colder and more saline ones during summer [35].

3.1 Marine sources of plastic pollution

3.1.1 Plastic products in aquaculture and fishery

Across the Mediterranean, but in the Adriatic Sea in particular, there is a continued demand to increase aquaculture production to fulfill the increasing market demand. Mussels, clams, sea bass and seabream production has become a significant source of regional income. Aquaculture was developed to support consumers' demand for seafood and the methods of production have continued to expand with the growing consumer market. As the need for fish and mussel aquaculture has increased, the development and expansion of aquaculture facilities in coastal and open water locations has increased accordingly. The expansion of the industry and the diversity of materials used to build and maintain aquaculture systems have paralleled the development of synthetic polymers over recent decades. Synthetic fibers offer greater strength and durability than natural fiber ropes; they are cheap, durable and easier to handle compared to their natural counterparts. Most modern aquaculture activities use plastic-based lines, cages, or nets suspended from buoyant or submergible structures (in part made of plastic) and have nanotech plastic-based biofouling and paint applied. Today, tanks, pens, nets, floats, pontoons as well as the pipes of the fish feed supplying systems are made of plastic materials. All plastic material within an aquaculture site is maintained and controlled for chemical degradation, biofouling and corrosion, and is regularly inspected to ensure strength and stability. In the context of global plastic pollution to the oceans, aquaculture may be a contributor to this. However, the estimation of their contribution remains a knowledge gap and lost or derelict gear as well as other possible plastics emissions from aquaculture can be a locally important contributor especially in coastal areas with intensive activity. New reports also point out a potential micro and nanoplastic contamination in wild and cultured seafood products even if the extent of such phenomena is still unknown. There is also concern regarding fisheries as a source of microplastics to the marine environment because both sectors use plastics that may degrade/fragment into microplastics. The coastal areas of Emilia Romagna and the Croatian coast represent sites of intense mussel and fish aquaculture production with hundreds of tons produced yearly. On the other hand, intense fishing activities coexist with a variety of fishing gear and methods being used in industrial and small-scale fisheries. Fishing gear for capture fisheries includes trawl nets, dredges, surrounding nets, lift nets, seine nets, traps, hook and lines. Nets and floats are made from a range of plastics including PP, PET, NyL, PVC, polyamide (PA) and PS.

3.1.2 Offshore oil and gas production activities

In oil and gas exploration, drilling fluids based on plastic microbeads were introduced a decade ago. Teflon strengthened particles have been largely applied for drilling purposes internationally. Despite the use of Teflon and other polymers with specific features being used extensively in production, waste treatment processes are not designed for, and give no mention of how to handle plastic particles, so this has clearly not been addressed as an issue in the past. Therefore, there is a substantial lack of information on potential loadings of microplastics used in this sector. To date, few fragmentary studies have addressed this topic. CEFAS's report entitled, "The discharge of plastic materials during offshore oil and gas operations" suggests that 532 tons of plastics and 7475 tons of "possible plastics" have been released from the UK offshore oil sector. Although knowledge about microplastic from oil and gas extraction activities is limited, it is very likely they represent a potential contributor in the emissions of plastics in aquatic environments, including microplastic and fibers, emphasizing that it should certainly be considered in future source assessments. The mapping of the distribution of rigs and platforms in the Adriatic Sea where tens of oil fields with hundreds of medium sized oil rigs occur, may provide estimations about the geographic distribution of the potential input related to these industrial activities.

3.1.3 Decommissioning of ships and oil rigs

Ships and maritime installations contain many plastic items, like insulation, coating, electrical wiring, furniture and textiles. Ideally, installations should be stripped of all potentially hazardous materials before dismantling. However, plastics items are not identified in the list of harmful materials. Therefore, polymer-based coatings and several kinds of insulation and wiring are rarely stripped.

3.1.4 Transportation and logistics

The distribution of products can contribute to the release of plastics in the environment. Most transferring of stock will occur alongside the transport infrastructure network. However, even if recognized as an important source of pollution, the contribution from releases during transportation, and as is the case for shipping, a map of the main transportation network including roads and harbors is still lacking. Systematic mapping in the Adriatic context has been suggested to improve the understanding of the areas where potential inputs can occur, providing a proxy for the potential intensity for release. The Adriatic Ship Traffic Database also contains information on ports in the Adriatic Sea that could be used to gauge the intensity of port activity to identify which of the port areas could potentially be receiving the largest inputs. Furthermore, the cruise ship industry is pointed out as a significant contributor to the problem of plastic pollution in the Adriatic sea. However, very limited data are available and no specific regulations in place for their plastic waste management and/or assessment of their environmental impact [36].

3.2 Land-based sources

3.2.1 Waste management

At a global level, the major challenge to tackle the input of plastic debris from land into the ocean is the lack of adequate waste management in coastal regions with a high and growing population density. Due to a generally high population density in coastal areas of the Adriatic, the pressure resulting from land-based inputs should be relatively high overall. Given such levels of anthropogenic pressure, the lack of, or deficient local waste management systems may lead to locally high inputs linked to industrial or domestic waste management.

There are no studies looking specifically at the leakage and marine input of plastic debris linked to these waste management systems, but ongoing work to quantify and characterize beach litter here points toward potential input from inadequate waste management on the eastern shores of Croatia where the islands of the Quarnero natural park present high loadings of plastic fragments. The composition of the waste accumulated resembles the composition of surveys carried out in the mid-Adriatic region where influence from higher population densities along the coastline is being registered. In addition, a study looking into microplastics near Venice has detected exceptionally high concentrations of small plastic fragments and microplastics in a nearby sandy beach [52]. Though not specified in this report, this exceptionally high concentration of microplastics, including large amounts of plastic fibers and film, could be linked to this location being close to the harbor as well as the lack of waste management facilities. To gain further insight into the potential release of plastics associated with waste management, it would be useful to map the distribution of population density as well as the location of urban agglomerations and settlements as this information will provide an indication of potential localized points of release of plastic waste into the environment. This kind of information is readily available at a sufficient resolution to allow identification of the areas within the Adriatic Sea that need more attention to this potential source of plastic pollution.

3.2.2 Sewage treatment plants

A rough estimation predicts that 70–80% of marine litter, composed primarily of plastics, originate from inland sources, ending in rivers and oceans. However, inland deposition of MP has not been investigated thoroughly. Potential sources include sewage treatment plants (STPs) and runoff from urban, agricultural, tourist, and industrial areas. As the retention capacity of conventional wastewater treatment processes to MPs appears to be variable in both magnitude and specificity, a characterization of MP emission by STPs and other sources is needed to map major sources of freshwater and terrestrial MPs. A relevant input to the terrestrial ecosystem is by fertilizers obtained by processing sewage sludge, as it typically contains more MPs than liquid effluents. Such fertilizers are frequently used in agriculture, implying a potential accumulation of plastic particles in the soil with continued use, and a systematic examination and quantification has been addressed by several research groups around the world. However, due to runoff, deposited plastic items are most likely transported to rivers and other waterways and ultimately discharged into estuarine and marine environments.

3.2.3 Agricultural production

The north of Italy and Croatia represent areas of intense horticultural activities where the agricultural practice of plastic mulching is prevalent. Plastic sheets are used to cover soil in order to preserve moisture, improve fertility and reduce weed infestation. Very often, fragments of plastic films are left behind after use and may accumulate in the soil, further fragmenting to produce nanometric particles. It has been estimated that 125–850 tons of microplastic per million inhabitants are added each year to agricultural soils in Europe, with an annual total of 63,000–430,000 tons of microplastic added to European farmlands. The northern part of Italy and Croatia is an area of significant agricultural and horticultural activities, therefore representing a potential hot spot for the release of plastic fragments in the terrestrial ecosystem. However, due to runoff phenomena these plastic items are most likely transported to rivers and other waterways and ultimately discharged into the estuarine and marine environments.

3.2.4 City dust and road wear

The first pilot studies of microplastic abundance in confined areas of heavily populated areas like the Oslo fjord noted that a large fraction of particles may be related to city dust (e.g. asphalt and car tires). City dust in urban runoff is known as a significant source of pollution to waterways. Plastics, such as styrene-butadiene, styrene-ethylene-butylene-styrene copolymer, are also used in road materials to make the asphalt more elastic [37]. Another potential contributor to the emissions of plastic fragments is road marking paint as these paints have a variable fraction (1–10%) of thermoplastic component (e.g. styrene-isoprene-styrene, ethylene-vinyl acetate, polyamide and acryl-monomer). On the other hand, the tread of car tires is largely based on styrene-butadiene rubber, a synthetic polymer formulation. Therefore, road dust entering the sea through air or storm water carries a significant fraction of microplastic from road materials, marking paint and car tires.

3.3 Pathways and distribution

The description and understanding of the pathways of the entry of marine plastic pollution into the Adriatic Sea is a central element in tracing the pollution back to its sources and developing effective plastic pollution preventing policies. A complete understanding of the input of plastic pollution into the aquatic environment needs to consider the source sectors and the mechanisms of transportation, distribution and partition through different environmental matrices. If the release occurs in the terrestrial environment, rivers and wind or atmospheric circulation constitute the logic pathways. When considering the presence of plastic debris and microplastics in a part of the global Mediterranean Sea there is a need to consider the transfer of marine plastic pollution into the relevant part of the large water bodies through the regional circulation pathway like the Adriatic Sea. The understanding of the input through these pathways is crucial in gauging the relative importance of local sea-based or coastal sources versus remote sources within the Arctic watershed or from other parts of the ocean.

3.3.1 Riverine input

The Adriatic Sea has a limited watershed. The largest rivers in the area are mostly located in the northern sector and include the Po, Adige, Tagliamento, and Arsa rivers. In terms of discharge, the Po River has the largest discharge with 1540 m^3/s followed closely by the Adige River with 235 m^3/s. The Po Basin is home to some 14 million people and extends over 24% of Italy's territory. The Po catchment is densely populated and subjected to high anthropogenic pressure heavily anthropized. Indeed, it represents the largest cultivated area in Italy and accounts for one third of national's agricultural production. The area account also for one of the highest concentrations of economic activities. Such massive river discharges make terrestrial influences particularly strong in the Adriatic Sea. However, to date there is no monitoring of the flux of plastics from rivers into the Adriatic Sea and though it has been identified as a possible pathway, the contribution of riverine discharge to plastic input is expected to be high because these rivers flow through densely populated and anthropized watersheds.

3.3.2 Atmospheric input

It has been speculated that at the global level much less plastic debris is transported by wind than by rivers [38, 39]. However, wind transport of plastic debris may be significant, particularly in coastal areas dominated by strong periodic winds. Wind may be a significant contributor in lightweight debris distribution. During intense storms wind can mobilize debris that would not normally be available for transport and carry it directly into rivers and the sea. Wind-blown litter is likely to be considerable as the Adriatic Sea is characterized by periodically windy shorelines. Atmospheric circulation has been proven to provide an efficient pathway for the transportation of floating microfibers and small plastic particles in the Mediterranean Sea as well as in other areas [33, 40]. Furthermore, some preliminary transport models tailored to the Adriatic oceanographic conditions, considering the contribution of waves and wind in the surface plastic distribution, define the Adriatic Sea as a highly "dissipative" system with respect to floating plastics with a calculated half-life of floating condition of 43.1 days [41, 42]. The authors conclude by pointing out that by construction the Adriatic coastline may be responsible for the main sink of floating plastic debris.

3.3.3 Oceanic input

The contribution of inputs through the movement of marine water masses by currents also needs to be considered in the global distribution model. The Adriatic region is poorly connected to the Mediterranean through the southern edges of the Otranto strait and the Ionian Sea exchanging with the Mediterranean Sea. The exchange of water, and possibly any moving plastic pollution, from and to the Mediterranean Sea has recently been addressed by the modeling work of Liubartseva et al. [40] and partially by the results of Pasquini et al., [40] which pointed out the formation of an accumulation zone corresponding to the three well known gyres located northside, central and in the southern sector of the Adriatic Sea.

4. Occurrence of plastic litter in the Adriatic Sea

4.1 Levels of macro- and microlitter in beaches

Some key research projects have recently addressed the need of defining the baseline levels of litter (macro-, meso- and microplastics) in the intertidal areas of beaches within the Adriatic Sea. Blašković et al. [41] investigated the occurrence of plastic debris in several sites of the Natural Park of Telaščica (Croatia). In all analyzed sites, fibers were the most recurring shape (90%) within the identified plastic debris while films where the second most common plastic fragment observed (7%) followed by pellet, foams, granules and unrecognized plastic pieces. Most of the plastic debris belonged to the size fraction from 1 mm and 64 μm (88%) followed by the fraction between 1 and 2 mm (11%). These results confirm previous characterization efforts of Laglbauer et al. [43] in six Slovenian beaches located in the gulf of Trieste (North-East Adriatic Sea). Within this assessment the authors sorted out a total of 5870 macro-debris units, yielding a median density of 1.25 items/m^2. The detailed analyses of the processed samples revealed a dominant secondary microplastics source being fibers the 85% of the total observed plastics and a number of 155 particles m^2 in the infralittoral zone, and 133 particles m^2 on the shoreline. On the Adriatic beaches surveyed, plastic dominated in terms of abundance, followed by paper and other groups. The average density was 0.2 litter items m^2, but at one beach it raised to 0.57 items m^2. Among plastic, cigarette butts were the most frequently found type of litter, and other plastic items with the highest occurrence were: small fragments, bottles and bottle caps, cutlery, and mesh bags. Their presence is a good indicator of pollution from beach users [44]. Most of the beached marine litter are from land-based sources, but with different sources and contributors. The main source of litter was primarily touristic activities, accounting for 37.9% of found litter which is lower than r the Mediterranean average (52%; [45, 46]). Filter cigarette were the second litter origin, but with a value (25.5%) lower than indicated for the Mediterranean (40%) [44]. The high percentages of in situ deposited litter found in the investigated sites are caused by the high number of visitors, more than 700,000 annually mainly during the touristic season (see i.e., http://statistica.regione.veneto.it; http://imprese.regione.emilia-romagna.it).

4.2 Levels of macro- and microlitter in surface waters

Few studies have addressed the occurrence of floating plastic debris in the surface water of the Adriatic Sea. Suaria et al. [33] reported by a larger study addressing the Mediterranean Sea and partially the Adriatic sector a clear prevalence of smaller particles. Quantitative estimations collected by a 400 μm net mesh pointed out values ranging from 0.4 ± 0.7 to 1.0 ± 1.8 items/m^3. The overall result the study pointed out that, within a total no. of 14,106 scored particles, 26% of all counted particles were smaller than 300 μm while 51% were smaller than 500 μm being the mean abundance of these meso-particles of 0.016 ± 0.028 particles/m^2. PE was the predominant form with an overall frequency of 52%, followed by PP (16%) and synthetic paints (7.7%). Polyamides (PA) accounted for 4.7% of all categorized particles which accounted alone for 2%), while PVC, PS and PVA represented equally contributed with 3% of the total. Other less frequent polymers (<1%) included: PET, polyisoprene, poly(vinyl stearate) (PVS), ethylene-vinyl acetate (EVA) and cellulose acetate. Noteworthy the authors concluded that the composition of western Mediterranean samples was dominated by low-density polymers such as polyethylene and polypropylene while the processed Adriatic samples instead were more heterogeneous and rather characterized by a higher presence of paint chips, PS, PVC, PVA and PAs. Within the "Derelict Fishing Gear Management System project – "DeFishGear" project co-funded by IPA-Adriatic Cross-border Cooperation Programme and the European Union, 120 visuals transect surveys were conducted during three cruises, covering a total length of 922.2 km [47]. A total of 1364 macro marine debris objects were observed floating on the Adriatic. The densities of the recorded floating debris were 5.66 items/km^2. The authors estimated that the observed floating marine debris was mostly originated from coastal segments close the high-density population cities and major rivers and transported by cyclonic surface circulation until either stranding. They calculated an average time from source to the sighting point of 22.8 days. These outcomes support Carlson and co-workers [48] previous assessment where an average residence time of 22.9 days but with also an average transit times of 20–60 days from a coastal region in the northwest Adriatic to a coastal region in the southwest [47]. The transport pathways, residence times, and probable sources and sinks identified further support with previous studies of the Adriatic Sea surface circulation and marine debris published by Liubartseva et al., [40].

4.3 Levels of macro- and microlitter in sediments

Data regarding macro- and mesolitter on the sea-floor in the Adriatic Sea are also available from the "SoleMon" Project (Solea Monitoring—Rapido trawl survey in the Northern Adriatic Sea), carried out since 2005 in the Northern and Central Adriatic Sea [49]. Plastic litter was divided by the authors in three sub-categories based on its source: fishing nets, mussel culture debris and other plastic e.g., bottles, plastic glasses, bags. Lost fishing nets and mussel culture debris accounted for 50% of the overall plastic litter collected over the investigated period. The remaining plastic comprised a wide range of objects such as garbage bags, shopping bags, cups, bottles, food packaging, dishes, other kitchen stuffs and industrial packaging [40, 48]. Results of this study indicated that the largest amount of mussel culture debris was found close to the coast and its distribution was constant over the years. These nets might have been accidentally lost/abandoned at sea during the collection and preparation of the product [50]. In the meantime, the fishing nets were found mainly close to the coast within 3 nm. This distribution was explained as fishing nets were mainly set-nets used by small scale fisheries that usually fish not further than 3 nm where there is not trawl fishing that can destroy these nets. A significant contribution of plastic litter found close to the coast was represented by food packaging, plastic bags, bottles and dishes or kitchen tools. The land origin is due to the municipal solid waste [48]. The authors concluded considering that the distribution varied among the years, but the occurrence was mostly related to both the close position of the sampling site to large cities along the coast, where the population density increases during the touristic season as well as the contribution of river [40, 50, 51]. As regards the microliter in the sedimentary environment, a preliminary assessment of microplastics in marine sediments along a coast- off-shore transect in the Central Adriatic was performed by Munari et al. [44]. Plastic fragments recollected from 64 samples were scored, weighted and identified by FTIR. Microplastics ranging 1–30 mm were found in all analyzed samples. The most recurring shapes were filaments-like (69.3%), followed by fragments-like (16.4%), and film-like (14.3%). In term of size distribution, plastic fragments in a range from 1 to 5 mm accounted for 65.1% of debris, while larger fragments (5–20 mm) contributed with the 30.3% of total amount, while larger fragments >20 mm represented the 4.6% of total. Six were the most recurring polymer types: nylon, polyethylene and ethylene vinyl alcohol copolymer. Furthermore, sediments from several sampling sites located in Italy, Slovenia, Croatia, and Greece were also analyzed for plastic debris content by the "DeFishGear" project. Plastic fragments in beach sediments were ranked into large sized particles (1–5 mm) and small microplastic particles (<1 mm). In general, microplastic from 1 to 5 mm ranged from 11 to 710 items/m^2. On the other hand, the fraction of smaller size scored from 70 to 6724 items/kg of dry sediments. The mean concentration for all Adriatic region was calculated as 113 ± 101 items/kg for the larger sized fragments and 1133 ± 1271 items/kg of dry sediments for the smaller ones. In detail, the selected Croatian beaches showed considerably greater presence of smaller microplastic per kg of sediment with value of approx. 227 items/kg of sediment while the larger sized fragments sored values approx. Ten times lower (17–28 items/kg of dry sediments). The composition of sorted fragments <1 mm showed the prevalence of plastic fragments as fragments represented approx. 70% of the total while filaments represented the left 29% of the total while a limited amount (1.8 and 0.9%) were film and foams. The chemical characterization of microplastic of the larger particles was performed on foams, pellets, fragments and filaments, while filaments and films were analyzed among the smaller sized particles. Beside the PE and PP in a few percent also PA, PET, PES, PS, PO, nylon and acrylic fibers were present among larger particles, while among the smaller viscose was detected. In the Greek sector data were obtained from three sites: the Halikounas, Issos and Acharavi beaches. The mean concentration of 1–5 mm sized debris varied from 68 items/m^2 (Halikounas) to 58 items/m^2 (Acharavi) while the small sized fraction of Ø > 1 mm showed values from 19 to 7 items/m^2 respectively for Halikounas and Acharavi. The most abundant categories on Halikounas beach were fragments and foam, while on the contrary pellets were the most abundant in Issos and Acharavi beaches. Chemical characterization of fragments, for Halikounas beach were done being both PE and PP the most recurring polymers in the larger particles while PP was the most occurring polymer in the smaller size fraction. The same project also addressed the occurrence in the Italian sector. High amount of small microplastic particles (<1 mm), up to 2526 items/kg of sediment, was found in the Cesenatico area. In the meantime, a limited amount corresponding to 0.56–1.02 items/kg of large particles (1–5 mm) were reported. Overall, 73% of the small microplastic particles were characterized by fragments while the remaining 26% as filaments. On the other hand, the large microplastic particles had different amount of all categories; however, fragments resulted the most abundant category (44%). The chemical identification showed PE as the most abundant material, followed by PP, PO, PES, PS and PAN. In the Slovenian coastline the selected sampling site showed a higher abundance of small microplastic particles (615 items/kg) respect of large microplastic particles (516 items/kg). In detail, the analysis of the small size fraction reported filaments being the predominant type of the microplastic composition, with representation of approx., 76% of the total. The second most common type of microplastic category were fragments and the third were films, with occurrence high as 9.5%. The chemical identification pointed out PE as the most recurring polymer type in the analyzed sediment samples, followed by PP, PET and PVC. Finally, Vianello and co-workers investigated the Venice Lagoon, a fragile estuarine ecosystem dominated by diversified anthropogenic activities, suspected to be a hot spot of plastic debris contamination [53]. Plastic debris of ≤1 mm or less was investigated in sediments collected from 10 sites chosen in shallow areas. Total abundances of plastic fragments varied from 2175 to 672 items/kg with higher concentrations generally found in the inner parts of the Lagoon. PE, PP, ethylene propylene (PEP), polyester (PEst), polyacrylonitrile (PAN), PS, alkyd resin (Alkyd), PVC, polyvinyl alcohol (PVOH) and NyL were identified. PE and PP were the most recurring polymer in the investigated samples which accounted for more than 82% of the total detected plastic debris in the whole sampling area. Among all classified shapes, irregular fragments accounted of the 87% of the total while films (2%) and pellets/granules (1%) were only occasionally recognized [54].

4.4 Levels of microliter in biota

The first report on the harmful effects of plastic debris ingestion on marine species in the Adriatic Sea was published in 1999 [55]. A dead dolphin *S. coeruleoalba* with the stomach occluded by different kinds of plastic materials was found near the island Krk, in the North Adriatic Sea. A following study on the logger head sea turtles, *C. caretta*, revealed a percentage of 35.2% of turtles sampled in the eastern Adriatic Sea were affected by plastic debris [55]. Occurrence of MPs in the gastrointestinal tract and gills of pelagic and demersal fish and marine mammals have been reported [56]. Few plastic debris accumulation studies have been performed in the Adriatic Sea. Pellini et al. [57] aimed at characterizing the occurrence, amount, typology of microplastic litter in the gastrointestinal tract of a benthic fish, *S. solea*, in the northern and central Adriatic Sea. The digestive tract contents of over 500 individuals were collected from 60 sampling sites and examined for microplastics. These were recorded in 95% of sampled fish, with more than one microplastic item found in around 80% of the examined specimens. The most commonly found polymers were PVC, PP, PE, polyester (PES) and PA. In details, 72% of the total classified plastic debris were fragments and 28% were identified as fibers. The mean number of ingested microplastics was 1.6–1.7 items/fish. PVC and PA showed the highest densities in the northern Adriatic Sea, both inshore and off-shore while PE, PP and PET were more concentrated in coastal areas with the highest values offshore from the port of Rimini. These results confirm previous observations of Avio and co-workers [13] in various fish species collected along the Adriatic Sea. FTIR analyses indicated PE as the predominant polymer (65%) in the stomach of fish. More than 100 fish representatives of five commercial species like *S. pilchardus, S. acanthias, M. merlucius, M. barbatus C. lucernus* were collected from the Central and North Adriatic Sea. The mean number of ingested microplastics was 1.0–1.7 items/fish. In details, the shape of the plastic debris observed in the stomachs of the investigated samples was mostly fragments and line followed by film and pellet. The 18% of extracted microplastics exhibited the larger size class (from 5 to 1 mm), 43% was between 1 and 0.5 mm, 23% between 0.5 and 0.1 mm, and the 16% lower than 0.1 mm. The chemical characterization pointed out that approximately 65% of analyzed plastic fragments were PE, followed by PET, PS, PVC, Nylon and PP. These early findings suggest the possible accumulation of plastic debris through the food web. Despite of some recent findings point out that at the bottom of the food pyramid, filter feeders, such as mussels can ingest and incorporate MPs in their tissues [58], more research is needed to unveil the abundance, distribution and polymeric composition of plastic debris in marine organisms at different levels ecological web in areas like the Adriatic Sea were multiple anthropogenic activities coexist.

5. Conclusions

The few available studies in the area prove the ubiquity of plastic pollution in the Adriatic Sea. The peculiar oceanographic conditions as well as the high levels of plastic debris recorded in all investigated matrices tend to classify such enclosed area as a hot spot of plastic contamination. Despite the distribution and circulation models appear to accurately estimate fluxes and final fate of marine plastic debris, sinks, sources, fate and residence times of different polymers at sea are the knowledge gaps that need to be addressed in the future to provide concrete info to support concrete actions toward plastic contamination reduction and remediation solutions.

Acknowledgements

The authors wish to thank The International Research Institute of Stavanger and the National Research Council of Italy- Institute of Marine Science for technical assistance and financial support to publish this work.

Conflict of interest

The authors declare no conflict of interest.

Author details

Alessio Gomiero[1,2*], Pierluigi Strafella[2] and Gianna Fabi[2]

1 NORCE Environment, Randaberg, Norway

2 National Research Council-Institute of Marine Sciences (CNR-ISMAR), Ancona, Italy

*Address all correspondence to: algo@norceresearch.no

References

[1] Plastics Europe. Annual Review 2017-2018. https://www.plasticseurope.org/en/resources/publications/498-plasticseurope-annual-review-2017-2018

[2] Wright SL, Thompson RC, Galloway TS. The physical impacts of microplastics on marine organisms: A review. Environmental Pollution. 2013;**178**:483-492

[3] Welden NA, Cowie PR. Degradation of common polymer ropes in a sublittoral marine environment. Marine Pollution Bulletin. 2017;**118**(1-2):248-253

[4] Gewert B, Plassmann MM, Mac Leod M. Pathways for degradation of plastic polymers floating in the marine environment. Environmental Science: Processes & Impacts. 2015;**17**(9):1513-1521

[5] Song YK, Hong SH, Jang M, Han GM, Jung SW, Shim WJ. Combined effects of UV exposure duration and mechanical abrasion on microplastic fragmentation by polymer type. Environmental Science & Technology. 2017;**51**(8):4368-4376

[6] Lobelle D, Cunliffe M. Early microbial biofilm formation on marine plastic debris. Marine Pollution Bulletin. 2011;**62**(1):197-200

[7] Zettler ER, Mincer TJ, Amaral-Zettler LA. Life in the "plastisphere": Microbial communities on plastic marine debris. Environmental Science & Technology. 2013;**47**(13):7137-7146

[8] Harrison JP, Hoellein TJ, Sapp M, Tagg AS, Ju-Nam Y, Ojeda JJ. Microplastic-associated biofilms: A comparison of freshwater and marine environments. In: Freshwater Microplastics. Cham: Springer; 2018. pp. 181-201

[9] Andrady AL. Microplastics in the marine environment. Marine Pollution Bulletin. 2011;**62**(8):1596-1605

[10] Engler RE. The complex interaction between marine debris and toxic chemicals in the ocean. Environmental Science & Technology. 2012;**46**(22):12302-12315

[11] Raynaud J. Valuing Plastics: The Business Case for Measuring, Managing and Disclosing Plastic Use in the Consumer Goods Industry. UNEP. 2014

[12] Browne MA, Niven SJ, Galloway TS, Rowland SJ, Thompson RC. Microplastic moves pollutants and additives to worms, reducing functions linked to health and biodiversity. Current Biology. 2013;**23**(23):2388-2392

[13] Avio CG, Gorbi S, Regoli F. Experimental development of a new protocol for extraction and characterization of microplastics in fish tissues: First observations in commercial species from Adriatic Sea. Marine Environmental Research. 2015;**111**:18-26

[14] Harmon SM. Chapter 8 - The effects of microplastic pollution on aquatic organisms. In: Zeng EY, editor. Microplastic Contamination in Aquatic Environments. Elsevier; 2018. pp. 249-270. ISBN: 9780128137475, https://doi.org/10.1016/B978-0-12-813747-5.00008-4

[15] Von Moos N, Burkhardt-Holm P, Köhler A. Uptake and effects of microplastics on cells and tissue of the blue mussel *Mytilus edulis* L. after an experimental exposure. Environmental Science & Technology. 2012;**46**(20):11327-11335

[16] Dantas DV, Barleta M, Ferreira da Costa M. The seasonal and spatial patterns of ingestion of polyfilament nylon fragments by estuarine

drums (Sciaenidae). Environmental Science and Pollution Research. 2012;**19**:600-606

[17] Sanchez W, Bender C, Porcher JM. Wild gudgeons (*Gobio gobio*) from French rivers are contaminated by microplastics: Preliminary study and first evidence. Environmental Research. 2014;**128**:98-100

[18] Gomiero A, Strafella P, Pellini G, Salvalaggio V, Fabi G. Comparative effects of ingested PVC micro particles with and without adsorbed Benzo (a) pyrene vs. spiked sediments on the cellular and sub cellular processes of the benthic organism *Hediste diversicolor*. Frontiers in Marine Science. 2018;**5**:99

[19] Pittura L, Avio CG, Giuliani ME, d'Errico G, Keiter SH, Cormier B, et al. Microplastics as vehicles of environmental PAHs to marine organisms: Combined chemical and physical hazards to the Mediterranean mussels, *Mytilus galloprovincialis*. Frontiers in Marine Science. 2018;**5**:103

[20] Rochman CM, Hoh E, Kurobe T, Teh SJ. Ingested plastic transfers hazardous chemicals to fish and induces hepatic stress. Scientific Reports. 2013;**3**:3263

[21] Bakir A, Rowland SJ, Thompson RC. Enhanced desorption of persistent organic pollutants from microplastics under simulated physiological conditions. Environmental Pollution. 2014;**185**:16-23

[22] Batel A, Linti F, Scherer M, Erdinger L, Braunbeck T. Transfer of benzo [a] pyrene from microplastics to *Artemia nauplii* and further to zebrafish via a trophic food web experiment: CYP1A induction and visual tracking of persistent organic pollutants. Environmental Toxicology and Chemistry. 2016;**35**(7):1656-1666

[23] Auta HS, Emenike CU, Fauziah SH. Distribution and importance of microplastics in the marine environment: A review of the sources, fate, effects, and potential solutions. Environment International. 2017;**102**:165-176

[24] Anbumani S, Kakkar P. Ecotoxicological effects of microplastics on biota: A review. Environmental Science and Pollution Research. 2018:1-24

[25] Riesbeck S, Gutow L, Saborowski R. (2017). Does microplastic induce oxidative stress in marine invertebrates? YOUMARES 8 - Oceans across boundaries: Learning from each other, Kiel, 13 Sep 2017 - 15 Sep 2017

[26] Imhof HK, Schmid J, Niessner R, Ivleva NP, Laforsch C. A novel, highly efficient method for the separation and quantification of plastic particles in sediments of aquatic environments. Limnology and Oceanography: Methods. 2012;**10**(7):524-537

[27] Hidalgo-Ruz V, Gutow L, Thompson RC, Thiel M. Microplastics in the marine environment: A review of the methods used for identification and quantification. Environmental Science & Technology. 2012;**46**(6):3060-3075

[28] Liebezeit G, Dubaish F. Microplastics in beaches of the east Frisian islands Spiekeroog and Kachelotplate. Bulletin of Environmental Contamination and Toxicology. 2012;**89**(1):213-217

[29] Dekiff JH, Remy D, Klasmeier J, Fries E. Occurrence and spatial distribution of microplastics in sediments from Norderney. Environmental Pollution. 2014;**186**:248-256

[30] Dehaut A, Cassone AL, Frere L, Hermabessiere L, Himber C, Rinnert E, et al. Microplastics in seafood: Benchmark protocol for their extraction and characterization. Environmental Pollution. 2016;**215**:223-233

[31] Nuelle MT, Dekiff JH, Remy D, Fries E. A new analytical approach for monitoring microplastics in marine sediments. Environmental Pollution. 2014;**184**:161-169

[32] Eriksen M, Lebreton LCM, Carson HS, Thiel M, Moore CJ, Borerro JC, et al. Plastic pollution in the World's oceans: More than 5 trillion plastic pieces weighing over 250,000 tons afloat at sea. PLoS One. 2014;**9**:1-15

[33] Suaria G, Avio CG, Mineo A, Lattin GL, Magaldi MG, Belmonte G, et al. The Mediterranean plastic soup: Synthetic polymers in Mediterranean surface waters. Scientific Reports. 2016;**6**:37551

[34] Fossi MC, Panti C, Guerranti C, Coppola D, Giannetti M, Marsili L, et al. Are baleen whales exposed to the threat of microplastics? A case study of the Mediterranean fin whale (Balaenoptera physalus). Marine Pollution Bulletin. 2012;**64**(11):2374-2379

[35] Artegiani A, Paschini E, Russo A, Bregant D, Raicich F, Pinardi N. The Adriatic Sea general circulation. Part II: Baroclinic circulation structure. Journal of Physical Oceanography. 1997;**27**(8):1515-1532

[36] Avio CG, Cardelli LR, Gorbi S, Pellegrini D, Regoli F. Microplastics pollution after the removal of the Costa Concordia wreck: First evidences from a biomonitoring case study. Environmental Pollution. 2017;**227**:207-214

[37] Sundt P, Schulze PE, Syversen F. Sources of Microplastic-Pollution to the Marine Environment. Mepex for the Norwegian Environment Agency; Report no: M-321|2015

[38] UNEP. Marine plastic debris and microplastics. In: Global Lessons and Research to Inspire Action and Guide Policy Change. Nairobi: United Nations Environment Programme; 2016

[39] Cai L, Wang J, Peng J, Tan Z, Zhan Z, Tan X, et al. Characteristic of microplastics in the atmospheric fallout from Dongguan city, China: Preliminary research and first evidence. Environmental Science and Pollution Research. 2017;**24**(32):24928-24935

[40] Liubartseva S, Coppini G, Lecci R, Creti S. Regional approach to modeling the transport of floating plastic debris in the Adriatic Sea. Marine Pollution Bulletin. 2016;**103**:115-127

[41] Pasquini G, Ronchi F, Strafella P, Scarcella G, Fortibuoni T. Seabed litter composition, distribution and sources in the Northern and Central Adriatic Sea (Mediterranean). Waste Management. 2016;**58**:41-51

[42] Blašković A, Fastelli P, Čižmek H, Guerranti C, Renzi M. Plastic litter in sediments from the Croatian marine protected area of the natural park of Telaščica bay (Adriatic Sea). Marine Pollution Bulletin. 2017;**114**(1):583-586

[43] Laglbauer BJ, Franco-Santos RM, Andreu-Cazenave M, Brunelli L, Papadatou M, Palatinus A, et al. Macrodebris and microplastics from beaches in Slovenia. Marine Pollution Bulletin. 2014;**89**(1-2):356-366

[44] Munari C, Scoponi M, Mistri M. Plastic debris in the Mediterranean Sea: Types, occurrence and distribution along Adriatic shorelines. Waste Management. 2017;**67**:385-391

[45] Conservacy O. A rising tide of ocean debris. 2009 Report. Washington, DC, USA; 2010

[46] PNUE/PAM/MEDPOL. Results of the assessment of the status of marine litter in the Mediterranean. In: Meeting

of MED POL Focal Points No. 334; 2009. 91 pp

[47] Vlachogianni T, Fortibuoni T, Ronchi F, Zeri C, Mazziotti C, Tutman P, et al. Marine litter on the beaches of the Adriatic and Ionian Seas: An assessment of their abundance, composition and sources. Marine Pollution Bulletin. 2018;**131**:745-756

[48] Carlson DF, Suaria G, Aliani S, Fredj E, Fortibuoni T, Griffa A, et al. Combining litter observations with a regional ocean model to identify sources and sinks of floating debris in a semi-enclosed basin: The Adriatic Sea. Frontiers in Marine Science. 2017;**4**:78

[49] Strafella P, Fabi G, Spagnolo A, Grati F, Polidori P, Punzo E, et al. Spatial pattern and weight of seabed marine litter in the northern and Central Adriatic Sea. Marine Pollution Bulletin. 2015;**91**:120-127

[50] Melli V, Angiolillo M, Ronchi F, Canese S, Giovanardi O, Querin S, et al. The first assessment of marine debris in a site of community importance in the North-Western Adriatic Sea (Mediterranean Sea). Marine Pollution Bulletin. 2017;**114**:821-830

[51] Galgani F, Barnes DKA, Deudero S, Fossi MC, Ghiglione JF, Hema T, et al. Marine litter in the Mediterranean and black seas. In: Executive Summary. CIESM Work. Monogr. 46. 2014. pp. 7-20

[52] Guerranti C, Cannas S, Scopetani C, Fastelli P, Cincinelli A, Renzi M. Plastic litter in aquatic environments of Maremma Regional Park (Tyrrhenian Sea, Italy): Contribution by the Ombrone river and levels in marine sediments. Marine Pollution Bulletin. 2017;**117**:366-370

[53] Vianello A, Boldrin A, Guerriero P, Moschino V, Rella R, Sturaro A, et al. Microplastic particles in sediments of lagoon of Venice, Italy: First observations on occurrence, spatial patterns and identification. Estuarine, Coastal and Shelf Science. 2013;**130**:54-61

[54] Vlachogianni TH, Anastasopoulou A, Fortibuoni T, Ronchi F, Zeri CH. Marine Litter Assessment in the Adriatic and Ionian Seas. IPA-Adriatic DeFishGear Project, MIO-ECSDE, HCMR and ISPRA. 2017. p. 168. ISBN: 978-960-6793-25-7

[55] Lazar B, Gračan R. Ingestion of marine debris by loggerhead sea turtles, *Caretta caretta*, in the Adriatic Sea. Marine Pollution Bulletin. 2011;**62**(1):43-47

[56] Dantas DV, Barletta M, Da Costa MF. The seasonal and spatial patterns of ingestion of polyfilament nylon fragments by estuarine drums (Sciaenidae). Environmental Science and Pollution Research. 2012;**19**(2):600-606

[57] Pellini G, Gomiero A, Fortibuoni T, Ferrà C, Grati F, Tassetti AN, et al. Characterization of microplastic litter in the gastrointestinal tract of *Solea solea* from the Adriatic Sea. Environmental Pollution. 2018;**234**:943-952

[58] Gomiero A, Strafella P, Maes T, Øysæd K-B, Fabi G. First record of the occurrence, composition of microplastic particles and fibers in native mussels collected from coastal and marine areas of the Northern and Central Adriatic Sea. In peer review

Chapter 2

Technological Approaches for the Reduction of Microplastic Pollution in Seawater Desalination Plants and for Sea Salt Extraction

Katrin Schuhen, Michael Toni Sturm and Adrian Frank Herbort

Abstract

An increasingly serious and widespread problem is the introduction of plastics into the water cycle. The poor degradability leads to the plastic waste remaining in water for a long time and over time it fragments into smaller and smaller plastic particles. Both the visible plastic parts and in particular their decomposition products and functionalized plastic particles are an enormous burden. Seawater desalination and sea salt extraction are highly dependent on the quality of the seawater in terms of process utilization and cost structures, i.e., on the level of pollution. Especially microparticles represent a significant potential for blocking the microfiltration membranes (pore size > 100 nm) in the pretreatment and the very costly reverse osmosis (RO) membranes (pore size > 5 nm). An innovative approach for the removal of microplastics from industrially used seawater combines a chemically induced agglomeration and a new technological implementation step. The particular challenge in removing the synthetic impurities is not only their small size but also their inert properties against most of the physical and chemical additives for flocculation. With an easy implementation to existing systems, an economic aspect and a strong impact on the maritime ecological balance will be expected.

Keywords: microplastics, desalination, sea salt extraction, reverse osmosis, filtration, agglomeration, add-on technology

1. Introduction

The oceans perform a vital function as a regulator of the climate and absorb 25% of the CO_2 generated [1]. Through the production and consumption of foodstuffs, industrial and consumer goods, humankind produces large quantities of waste, whereof a considerable proportion ends up in the ocean sooner or later. Three-quarters of the waste in the ocean consists of plastic. This plastic is a steadily growing problem, costs the lives of ten thousands of animals every year and can also endanger humans.

Over 80% of the plastic material entering the ocean annually originates from land-based sources, which correlates with the fact that half of the world population lives in coastal regions [2, 3]. Large plastic waste represents the main contribution, including everyday objects like drink bottles and other types of plastic packaging. An estimated 4.8–12.7 million tons enter here annually [2]. The remaining input comes from plastic, which is released at sea, mainly from fishing—for example, due to lost and discarded fishing gear, which is estimated at 0.6 tons a year [4]. About 94% of the large plastic parts, which end up in the sea, sink with time to the ocean floor. Today, an average of 70 kg of plastic can be found on every square kilometer of the ocean floor.

Approximately, 350–400 years can pass before the plastic is completely degraded. As it moves through the seas, the plastic changes. Through weather conditions and waves, for example, it disintegrates into smaller and smaller fragments and from the macroplastic, the so-called secondary microplastics are formed [5]. If it directly enters the environment, it is designated as primary microplastics.

By definition, microplastics are small, solid, and water-insoluble plastic particles under 5 mm in size. In the meantime, these particles can be found in all bodies of water. They could even be detected in the Arctic [6].

The input of primary microplastics is estimated at 0.8–2.5 tons a year [7]. This enters mainly through tire abrasion and textile fibers, which enter the wastewater through washing clothes and thus end up in the environment. But also the dust from the wear on road paint, microplastics used in personal care products, marine coatings, and lost plastic pellets are important entry routes (**Figure 1**).

In general, a strong correlation can be observed between the population density and the microplastics concentration [8, 9] .The proximity to densely populated regions and poor waste management lead to particularly high levels of contamination [10, 11]. Sewage treatment plants or plastics manufacturing respectively processing companies are also important point sources and can release high volumes of plastic and microplastics locally (**Figure 2**) [12, 13].

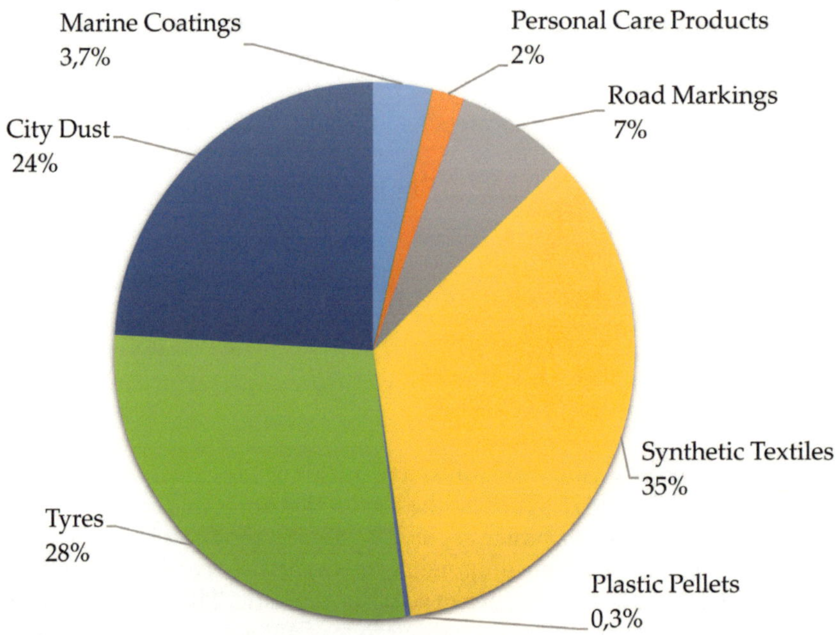

Figure 1.
Global release of primary microplastics to the world oceans [7].

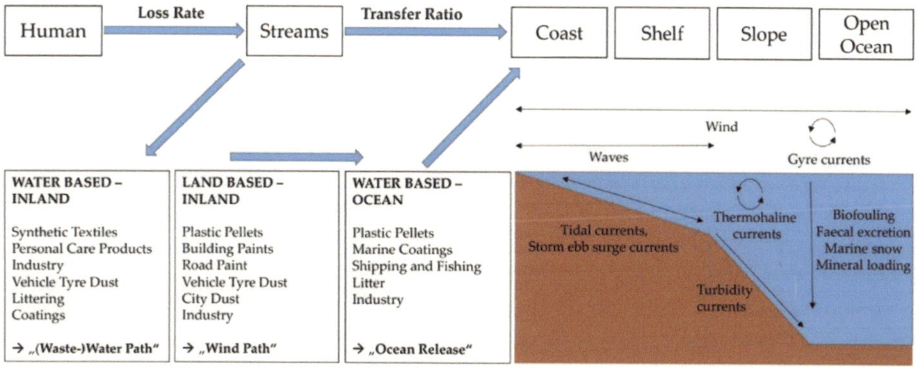

Figure 2.
Microplastic inputs and transport paths into the ocean [7, 15].

Ports and industrial areas are especially contaminated with microplastic particles [14]. The majority of the microplastics remain near the shore [10]. In the Arabian Gulf along the coast, 4.38×10^4–1.46×10^6 microplastic particles/km^2 could be detected in the surface water [15].

Off the coast of South Africa, there were 257.9 ± 53.36 to 1215 ± 276.7 microplastic particles/m^3 of water [12]. At the mouth of the Yangtze in the East China Sea, 4137.3 ± 2461.5 microplastic particles/m^3 could be detected, whereby the concentration in the open sea was only 0.167 ± 0.138 microplastic particles/m^3 [13].

The most common polymer types occurring in seawater are, in addition to polyethylene, polypropylene, and polystyrene, also polyamide, polyester, polymethylmethacrylate, polyvinyl chloride, polyoxymethylene, polyvinyl alcohol, polymethylacrylate, polyethylene terephthalate, alkyd resins, and polyurethane [16]. They are found in descending prevalence in the seawater.

The quantification of the inputs into the environment is, as a rule, based on a loss rate, which is calculated against the produced quantity of preproduced plastic [17]. The categorization of the industries, which manufacture preproduced plastic, is organized in producers (manufacture plastic material from raw materials), intermediaries, converters (convert preproduced plastic into products, or individual components), external waste disposers, and shipping companies (transport the material). By means of the difference between the respectively processed plastic quantities and the loss rate, the plastic quantities, which are released into the environment by the plastics industry alone, are revealed.

Forecasts assume an increasing plastic production volume in the future, which will lead to an increasing entry quantity of plastic and microplastics in the environment and seas [2]. In addition, microplastics arise continuously through the constant fragmentation of plastic already in the environment [5]. This leads to an ever higher contamination of the marine environment with microplastics.

Since the current analytical methods to detect microplastic in the aquatic environment have numerous shortcomings [18], the contamination of the marine environment can only be estimated. Particularly problematic here is that small plastic particles cannot be captured in most monitoring cases. The lower detection limit in the marine environment is usually 300 μm. With increased efforts, the lower detection limit can be reduced to 20–10 μm [16]. However, this is seldom practiced. Particles below the detection limit are not captured. Additionally, studies are not readily comparable, since there is no standardized monitoring procedure [18].

2. Seawater utilization

Seawater utilization can be divided in three primary use areas: the use for agriculture, for the home, and for industry [19]. While the principal portion is used for agriculture in developing countries, a significantly increased proportion for use in households and industry can be seen in industrialized countries. Advancing industrialization also causes increased water consumption. Over the last century, the world population has quadrupled, while the water demand has increased sevenfold.

In general, usable water comes from surface water, groundwater, or fossil aquifers. In order to increase the supply of usable water, processes such as the desalination of seawater have been moving into focus for some time [20]. Seawater has most commonly been used as a coolant in energy generation and in industrial processes so far. It is also used in mining to extract minerals as well as for the hydraulic fracturing of gas and oil. It is additionally applied in production processes, such as sea salt extraction, aquaculture, algal cultivation as well as food manufacturing [21]. It is also used for temperature moderation in buildings and areas as well as for cold water fishery (**Figure 3**).

Two characteristic economic seawater application sectors are seawater desalination and sea salt extraction. For sea salt extraction, seawater is diverted into large basins. Over time, the water evaporates by the heat of the sun and wind and the previously dissolved salt remains [22]. The media currently reports again and again about the contamination of sea salt with high quantities of microplastics. Thus, 50–280 microplastic particles/kg of salt were detected in Spanish sea salt and in Chinese sea salt 550–681 microplastic particles/kg of salt [23, 24].

Seawater desalination is the production of drinking water and process water for industrial facilities or power plants from seawater through the reduction of the salt content. The desalination can be based on various processes, which remove the salts and minerals from the water. To some extent, usable ancillary products like table salt accrue. In addition to the already present burden from anthropogenic stressors, the chemicals added to the seawater against fouling and scaling as well as the metals dissolved by corrosion threaten marine ecosystems, e.g., in the Red Sea. An

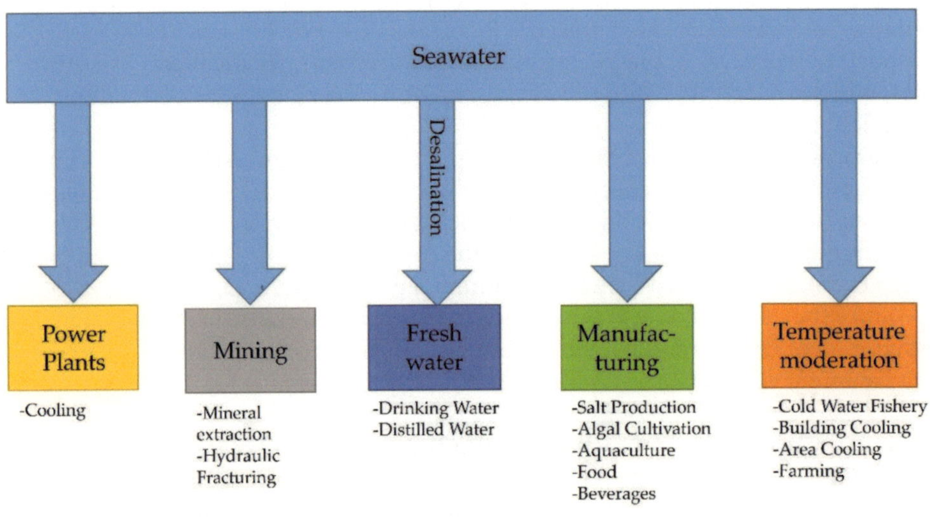

Figure 3.
Overview of the different seawater applications [21].

investigation of 21 plants, which together produce 1.5 million m^3/day, accounted for 2.7 kg of chlorides, 3.6 kg of copper, and 9.5 kg antiscaling agents piped into the sea per day of seawater desalination [25].

Seawater desalination is practiced in numerous coastal semiarid regions [20]. In addition to small plants in areas with insufficient infrastructure with only a few hundred cubic meters of water a day, there are also large desalination plants, e.g., in southern Europe (Barcelona Seawater, 200,000 m^3/day), the USA (Claude Bud Lewis Carlsbad, 204,000 m^3/day), Israel (Sorek, 624,000 m^3/day), Australia (Kurnell, 250,000 m^3/day), and the Unites Arab Emirates (Dschabal Ali Block M, >2,000,000 m^3/day). Significant expansion of production capacities for seawater desalination is, for example, planned in the Persian Gulf. Due to the development of the steel, petrochemical, cement, aluminum, and energy industry, there will be a demand of roughly 940,000 m^3/day only for use in these sectors until 2030 in Iran alone [26].

3. Impact of increasing microplastic burden on seawater utilization and the water use cycle

An increasing microplastic burden not only has implications for the costs and efficiency of the sea water utilization process, but also for the marine ecosystem and, as a consequence, also for humankind (**Figure 4**).

Bonded in microplastics, pollutants like, for example, dichlorodiphenyltrichloroethane (DDT), dioxins, or heavy metals can be transported and accumulated in organisms via ingested food [28]. Due to the manufacturing process, most polymer blends also contain harmful substances like softening agents or monomers, which in return can be released upon ingestion of the particles via food and exert a direct influence on the organism, since these substances are mostly classified as potentially harmful and/or carcinogenic [29]. It has already been shown in laboratory experiments that microplastics smaller than 150 μm can, after ingestion via food, enter the surrounding tissues, the bloodstream and, through these, the internal organs and also the brain [30]. There is then the risk of the formation of lesions and inflammations. Furthermore, oxidative stress, necrosis, and damage to DNA can be triggered, which again increases the risk of cancer. Neurological behavioral disorders are also possible [31]. Thus, there is a potential risk to human health from microplastics.

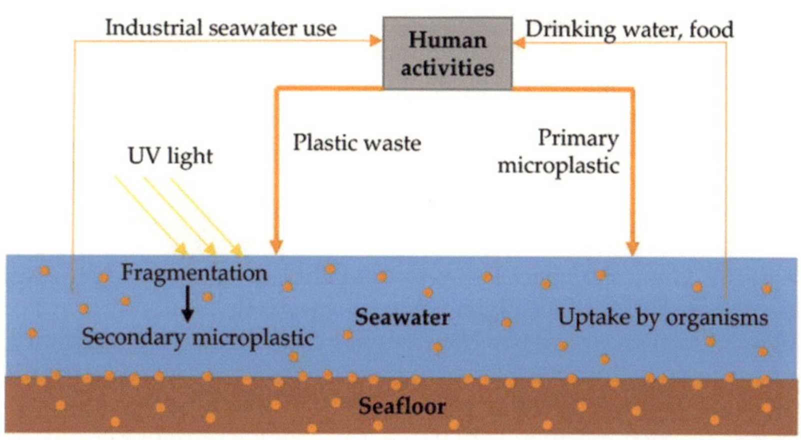

Figure 4.
Microplastics in the marine water cycle [27].

In addition to the decrease in improper disposal, the search for replacement substances and the prohibition of microplastics as a product addition for everyday products, ensuring that the aquiferous processes are free of microplastics also represents a chance to reduce the degree of contamination due to microplastics in the water cycle. Besides the implementation of new technologies for the purification of wastewater in sewage treatment plants, this also includes the conceptual and technical development of new add-on technologies in seawater utilization processes in order to filter microplastics out of the inflowing seawater and eliminate it prior to the seawater utilization processes.

4. Ecological-chemical approach to the reduction of the microplastic burden in seawater-based processes

At this time, there is no economical possibility yet to remove microplastics simply and cost-effectively from seawater. A promising research approach based on the adaptation of a concept by Herbort and Schuhen for freshwater systems and the simultaneous development of add-on technology for static (e.g., waterside plants) and mobile (e.g., ships) seawater utilization processes [32].

In the process developed by Herbort and Schuhen, silane-based microplastic agglomerates are formed according to the cloud point principle through the application of special organosilane-based precursors, which, via Van der Waals forces, have a high affinity to unreactive microplastics (IOCS, inert organic-chemical macromolecules) and, at the same time, a high reactivity in water. [32–35]. A Video shows the fixation process in a batch reactor for use in wastewater treatment [36].

Organosilanes are hybrid compounds of inorganic silanes and organic hydrocarbons [37]. Through the selection of the functional groups in the organic unit (functional design), it is possible to exploit an adaptable system for the respective application (e.g., removal of reactive and/or inert organic-chemical compounds). By means of the substituent pattern within the organic unit and also directly on the silicon atom, the affinity of the organosilanes can be adapted to various polymer types and, simultaneously, the reactivity respectively the stability can be controlled.

Organosilanes with corresponding reactivity can react to organic-inorganic hybrid silica gels in the sol-gel process [38, 39]. In the first step, disposal groups are split off and reactive silanol groups generated through hydrolysis. The silanol groups subsequently form silanone bonds in a condensation reaction and link the organosilanes via a bridging unit (**Figure 5**).

A three-dimensional networked hybrid silica gel forms, which is stabilized via the respective bridging unit. Through the selection of the disposal groups and the organic groups, the properties and reactivity of the organosilanes can be specifically controlled. The selection of the disposal groups is decisive for the hydrolyzability of the organosilanes [40, 41].

The organic groups influence the water solubility, the stability of the resulting silanols, and the affinity of the organosilanes respectively the ability of the resulting silanols to fix microplastic particles. By choosing specific organic groups, the affinity to different polymer types can be controlled. Groups with low polarity can be used to attach to polymers with lower polarity like polyethylene or polypropylene. More polar polymers containing, e.g., heteroatoms, like polyester or polyamide, can be fixed by using organic groups with a similar chemical structure and polarity.

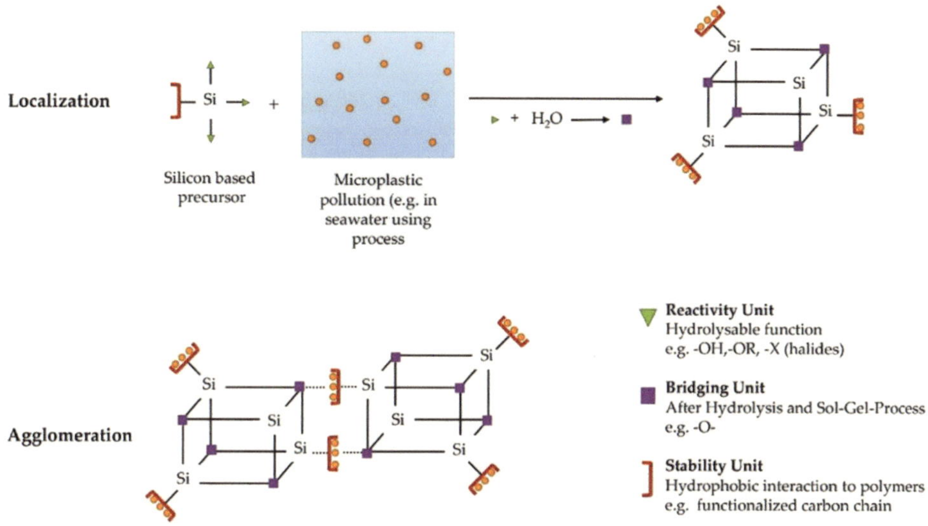

Figure 5.
Localization and agglomeration of microplastic particles [42].

The interaction of the disposal groups and organic basis units must be so coordinated that the highest possible affinity to microplastics and optimal reaction kinetics are achieved [33]. The reaction kinetics takes on a decisive function within the research approach. It is influenced, among other things, by the water matrix and the temperature. Knowing that an increase in the water temperature accelerates the speed of the sol-gel process and that the temperature of seawater is subject to seasonal fluctuations and spatial variations, challenges are presented for the implementation of the concept in the seawater environment [39]. In addition, the factors of salinity, temperature, UV radiation, and pH value play a determining role and are also to be considered (**Figure 6**) [43].

Dissolved salts influence reaction speed and the reaction mechanism, for example, through the electrolyte effect [44, 45]. It results in the stabilization of the intermediate stages with higher ionic strength as well as catalytic effects or competitive influences of dissolved ions. The pH value influences reaction kinetics through the catalytic effect of hydroxide and oxonium ions [39]. Temperature differences directly influence the reaction speed [46, 47]. Thus, a temperature increase of 10°C is accompanied, as a rule, by approximately double the reaction speed. UV light can also facilitate the formation of reactive intermediate stages via a photocatalytic effect and thus accelerate the reaction [46]. However, it can also break down the precursors used or already linked molecules.

Initial experiments at a laboratory scale show that the salinity has a slowing influence on the fixation process and the entire process of agglomeration formation also works in an artificial salt water matrix. To produce the salt water (3.5% salt by mass), 27.5 g NaCl, 5 g $MgCl_2$, 2 g $MgSO_4$, 1 g KCl, and 0.5 g $CaCl_2$ were dissolved in distilled water. This results in a mass concentration of 58.8% chloride (Cl^-), 29.6% sodium (Na^+), 4.7% sulfate (SO_4^{2-}), 4.9% magnesium (Mg^{2+}), 1.5% calcium (Ca^{2+}), and 0.5% potassium (K^+).

Subsequently, 0.1 g polyethylene powder (PE) (average particle size 350 μm), 0.1 g polypropylene powder (PP) (average particle size 350 μm), and 0.1 g of a 50:50 mixture of PE and PP were stirred in a beaker with 1 l of salt water respectively distilled water at room temperature. After 24 h, 0.15 ml agglomeration reagent was added and the mixture was stirred for an additional 24 h. The formed

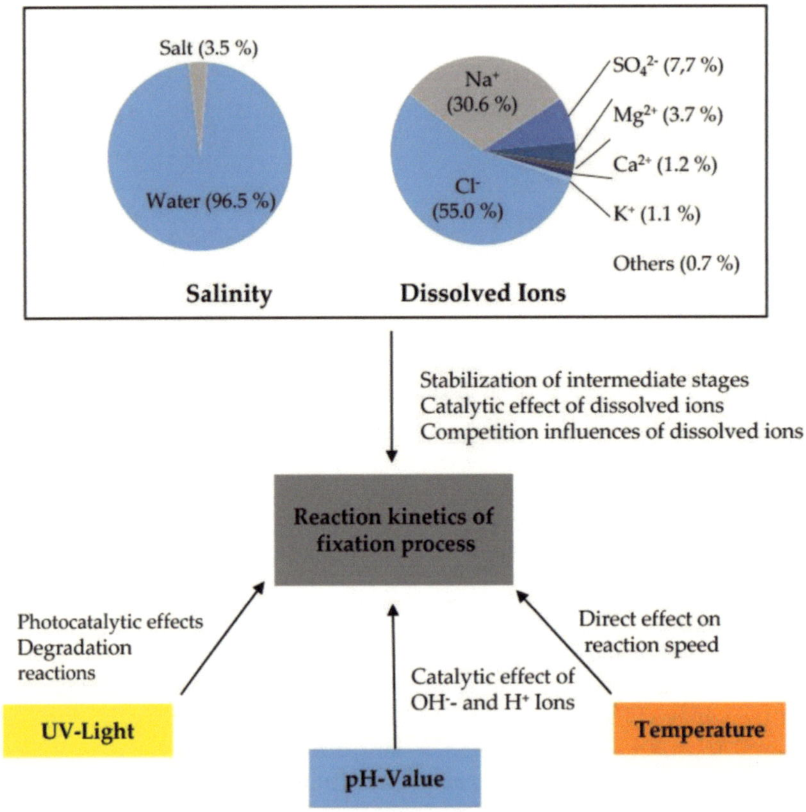

Figure 6.
Influence on the reaction kinetics of the fixation process in the marine environment (percentage by mass) [39, 43–47].

aggregates were removed and dried at 60°C for 24 h. ESEM images were taken using a FEI Quanta 250 ESEM (FEI Company, Hillsboro, USA) equipped with a large field detector (LFD). The chamber pressures were between 60 and 80 Pa and the acceleration voltage between 7 and 20 kV. The remaining water was filtered using a paper filter (Rotilabo 111A, 12–15 µm pore size). The tare weight of the filter was noted before filtration using a AX105DR (Mettler Toledo, Switzerland). Afterward, the filter was dried at 105° C for 24 h and weighted again, to check if there is remaining microplastic in the water (accuracy ±0.2 mg).

In distilled water, an aggregation of the microplastic particles begins 15 s after the addition of the agglomeration reagent. After 2–3 min, the agglomeration is completed and an aggregate is present, which contains all of the microplastics. In the artificially produced salt water samples, the agglomeration process begins after 10 min and is concluded after 15 min. It is, therefore, significantly slower, but nevertheless fixes all the microplastics. This shows that salt water has a stabilizing effect on the reactive intermediate stages and thus slows down the sol-gel process, whereby the agglomeration starts later and also takes longer. **Figure 7** shows ESEM images of the mircoplastic particles used and their aggregates formed during the fixation process respectively fragments of the aggregate prepared for the images. How the microplastic particles are linked and embedded by the agglomeration reagent can be observed, where a considerable increase in size results. As the agglomeration reagent reacts to a solid hybrid silica, which will be removed within the aggregates from the water, residues in the water will be avoided. To ensure a complete removal of the agglomeration reagent even in the trace substance range, further TOC analysis and particle analysis will be proceeded [48, 49].

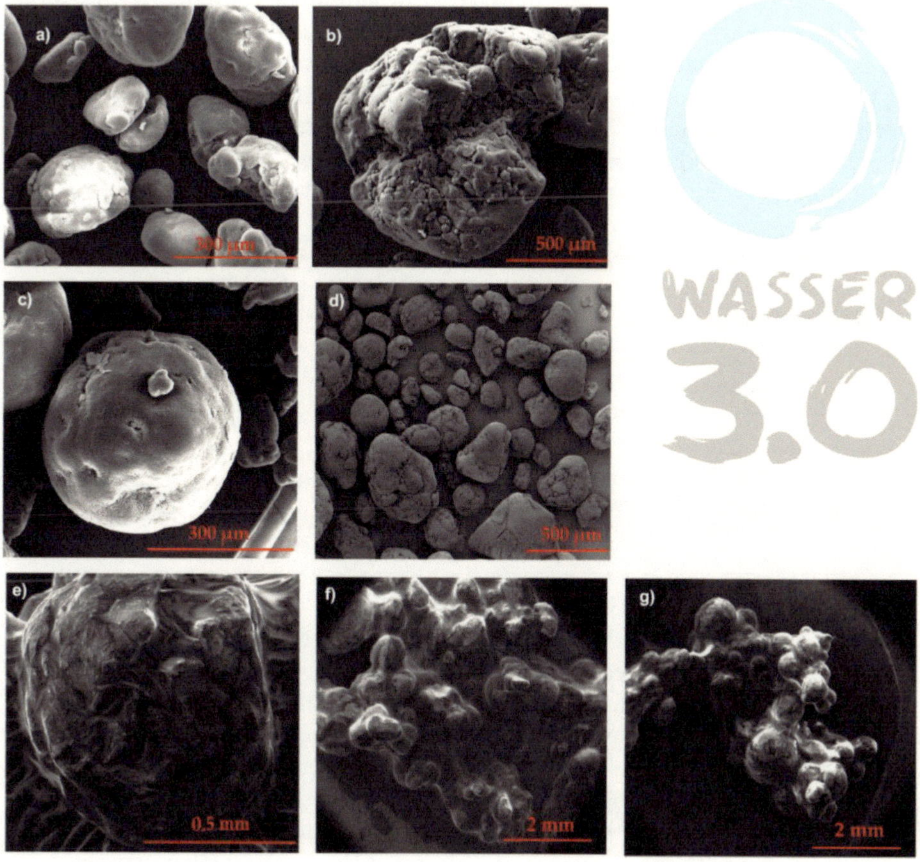

Figure 7.
ESEM images of the microplastic blanks (a, b = PE; c, d = PP) and of the agglomerates formed during the fixation process (e = PE; f = PP; g = PE/PP (50:50)).

5. Procedural implementation of microplastic elimination in seawater utilization processes

The procedural implementation pursues the goal of increasing the service life of the existing desalination plants by already holding back the microplastic particles (0.1–5 μm) initially in pretreatment, which cause blockages of the membranes (pore size 0.002–0.1 μm). On the one hand, the service life of the microporous membranes can be thus extended and the operational expenses reduced (without the addition of suspect additives, such as, e.g., antiscalants) as well as, on the other hand, sustainably eliminating the much-criticized microplastic particles from the water cycle. As the removal is based on a physicochemical agglomeration process, it is not limited by particles size or shape like a filtration process. The agglomeration reagent can bind to fibers, films, and fragments in all size classes and fix them in big agglomerates. As a consequence, the quality of the water on the removal side near the coast/surface as well as on the output side will be improved through the reduction of the microplastic particle load.

In the application in sea salt extraction, contamination of the resulting sea salt and thus the transmission to people will be effectively avoided through the removal of microplastics from the seawater flowing into the evaporation basins.

This is possible through the combined development of a pretreatment stage of a series of stirred tanks and the inorganic-organic functional material. In addition, a high-performance cascade process is developed as add-on technology in order to facilitate the material reaction and to make a throughput of >600 m^3 a day possible. Continuous operation is sought by connecting several cascades. The adapted concept as well as the related technological implementation strategy provides for the first time the opportunity to remove plastic particles with a particle size of <5 μm effectively and sustainably from salt water in an upstream, modular pretreatment step. The diagram of the process is presented in **Figure 8**.

The process is divided in the following stages:

- Inlet duct salt water/feed stream
- Dosing devices
- Mixing concept (in the reactor)
- Reactor
- Material discharge
- Outlet stream

The feed stream, that is the extracted seawater with unwanted microplastic load, is piped to the first partial reactor via an existing suction pipe. This should take place by means of upstream, abrasion, and corrosion-resistant pumps and use the suction pipes present in the existing seawater desalination plants. Depending on the load of the feed stream, a defined amount of the organosilanes is simultaneously added by means of the dosing device and blended with the salt water already in the first partial reactor through a mechanical mixing concept.

In accordance with the concept, the addition of the material takes place in several steps within the process. In this way, the required initialization period of the material and the reaction time can be responded to through the variable design, dimensioning, and number of dosing stages, mixing installations as well as partial reactors. For this purpose, können reactors of different sizes respectively different volumes can be used so that, for example, the reaction starts in the first partial reactor and continues to react optimally in the subsequent reactors. The organosilane is mixed with the salt water according to the required concentration in the respective partial reactors, at which time it is successively bonded to the microplastics to be removed. Through injection or chemical interaction, the concentration of free,

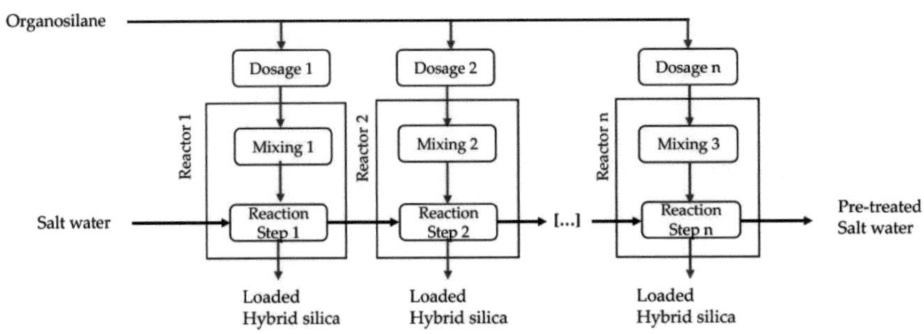

Figure 8.
Process flow chart.

Removal of microplastics from seawater (based on particle count)	≫85%
Removal of organics	Very high, >99%
Minimum size of removed particles (μm)	0.1
Use of additives	Low, product is an ecotoxicologically harmless agglomeration material
Reaction product (properties)	Particle load < 0.1 μm
Expected by-product (properties)	Loaded, inactive agglomeration material
Process and plant technology	
Process safety	High; in case of overdosing no ecotoxicologically negative effects
Plant availability	>95%
Plant requirements	Low, installation surface for container solution
Number of process steps	1 (agglomeration-flotation step)
Scalability	High, modularizable, and adaptable to degree of pollution

Table 1.
Technical target criteria.

nonbonded microplastic particles should continuously decrease along the length of the reactor respectively the residence time. A consistent residence time of the material in the reactor is needed so that a complete conversion can be surmised. This is essential for the process, since an incomplete conversion would mean additional burdens through further foreign substances/particles.

In order to prevent the disintegration of the agglomerates, the reactor has a strict and clearly defined residence time distribution (RTD) so that ideally all particles can pass through the reactor in the same ideal time frame and grow and that the disintegration of the agglomerates can be avoided. The implementation of microplastic elimination in seawater utilization plants pursues the technical target criteria presented in **Table 1**.

6. Conclusions

The innovative add-on technology for the removal of microplastics from industrial seawater utilization plants pursues the first problem solution regarding the risk of blocking from the immense microplastic particle load in the sea, among other things, in industrial, membrane-based seawater desalination plants (especially RO plants). Due to the significant reduction of the microplastic load in the pretreatment stage, alleviation results for the downstream RO membranes. The service life of the RO membranes will be significantly increased and the membranes can, thanks to the improved performance, be operated on a smaller scale and more cost-effectively.

Through the application of new add-on technology in sea salt extraction, the entry of potentially harmful microplastics in sea salt is reduced and thus, at the same time, also the contamination of everyday food. Based on the future increasing contamination of the oceans with microplastics, this technology helps to ensure the sustainable use of seawater.

Acknowledgements

The research projects of Wasser 3.0 (www.wasserdreinull.de) are conducted by means of the financial support by the German Federal Ministry for Economic Affairs and Energy through the provision of ZIM (Central Innovation Program for SME) project funds. The enterprise abcr GmbH (www.abcr.de) from Karlsruhe (Germany) and Zahnen Technik GmbH (www.zahnen-technik.de) from Arzfeld (Germany) are directly involved in the project as industrial partners for the material science scale-up as well as for plant construction and engineering. Michael Sturm thanks the German Federal Environmental Foundation (DBU) for the support with a PhD scholarships (Aktenzeichen: 80018/174). The authors thank Carolin Hiller, Lukas Pohl, and Laura O'Connell for their contributions.

Conflict of interest

We have no conflict of interest to declare. This research has not been submitted for publication nor has it been published in whole or in part elsewhere. We attest to the fact that all authors listed on the title page have contributed significantly to the work, have read the manuscript, attest to the validity and legitimacy of the data and its interpretation, and agree to its submission to the book.

Author details

Katrin Schuhen[1,2]*, Michael Toni Sturm[1,3,4] and Adrian Frank Herbort[1,3]

1 Wasser 3.0/abcr GmbH, Karlsruhe, Germany

2 Wasser 3.0/Zahnen Technik, Arzfeld, Germany

3 Institute for Environmental Sciences Landau, University of Koblenz–Landau, Germany

4 Water Chemistry and Water Technology, Karlsruhe Institute of Technology (KIT), Engler-Bunte-Institute, Karlsruhe, Germany

*Address all correspondence to: schuhen@wasserdreinull.de

References

[1] Bigg GR, Jickells TD, Liss PS, Osborn TJ. The role of the oceans in climate. International Journal of Climatology. 2003;**23**:1127-1159. DOI: 10.1002/joc.926

[2] Jambeck JR, Geyer R, Wilcox C, Siegler TR, Perryman M, Andrady A, et al. Marine pollution. Plastic waste inputs from land into the ocean. Science. 2015;**347**:768-771. DOI: 10.1126/science.1260352

[3] Bergmann M, Gutow L, Klages M, editors. Marine Anthropogenic Litter. Cham, Switzerland: Springer International Publishing; 2015. DOI 10.1007/978-3-319-16510-3

[4] Macfadyen G, Huntington T, Cappell R. Abandoned, Lost or Ortherwise Discarded Fishing Gear. Rome: Food and Agriculture Organization of the United Nations; 2009

[5] Barnes DKA, Galgani F, Thompson RC, Barlaz M. Accumulation and fragmentation of plastic debris in global environments. Philosophical Transactions of the Royal Society, B: Biological Sciences. 2009;**364**:1985-1998. DOI: 10.1098/rstb.2008.0205

[6] Peeken I, Primpke S, Beyer B, Gütermann J, Katlein C, Krumpen T, et al. Arctic Sea ice is an important temporal sink and means of transport for microplastic. Nature Communications. 2018;**9**:1505. DOI: 10.1038/s41467-018-03825-5

[7] Boucher J, Friot D. Primary Microplastics in the Oceans: A Global Evaluation of Sources. Gland, Switzerland: IUCN International Union for Conservation of Nature; 2017

[8] Driedger AGJ, Dürr HH, Mitchell K, van Cappellen P. Plastic debris in the laurentian great lakes: A review. Journal of Great Lakes Research. 2015;**41**:9-19. DOI: 10.1016/j.jglr.2014.12.020

[9] Wang W, Ndungu AW, Li Z, Wang J. Microplastics pollution in inland freshwaters of China: A case study in urban surface waters of Wuhan, China. Science of the Total Environment. 2016;**575**:1369-1374. DOI: 10.1016/j.scitotenv.2016.09.213

[10] Lebreton LC-M, Greer SD, Borrero JC. Numerical modelling of floating debris in the world's oceans. Marine Pollution Bulletin. 2012;**64**:653-661. DOI: 10.1016/j.marpolbul.2011.10.027

[11] Abayomi OA, Range P, Al-Ghouti MA, Obbard JP, Almeer SH, Ben-Hamadou R. Microplastics in coastal environments of the Arabian gulf. Marine Pollution Bulletin. 2017;**124**:181-188. DOI: 10.1016/j.marpolbul.2017.07.011

[12] Nel HA, Froneman PW. A quantitative analysis of microplastic pollution along the south-eastern coastline of South Africa. Marine Pollution Bulletin. 2015;**101**:274-279. DOI: 10.1016/j.marpolbul.2015.09.043

[13] Zhao S, Zhu L, Wang T, Li D. Suspended microplastics in the surface water of the yangtze estuary system, China: First observations on occurrence, distribution. Marine Pollution Bulletin. 2014;**86**:562-568. DOI: 10.1016/j.marpolbul.2014.06.032

[14] Claessens M, Meester S d, van Landuyt L, Clerck K d, Janssen CR. Occurrence and distribution of microplastics in marine sediments along the Belgian coast. Marine Pollution Bulletin. 2011;**62**:2199-2204. DOI: 10.1016/j.marpolbul.2011.06.030

[15] Zalasiewicz J, Waters CN, Ivar do Sul JA, Corcoran PL, Barnosky AD, Cearreta A, et al. The geological cycle of

plastics and their use as a stratigraphic indicator of the anthropocene. Anthropocene. 2016;**13**:4-17. DOI: 10.1016/j.ancene.2016.01.002

[16] Hidalgo-Ruz V, Gutow L, Thompson RC, Thiel M. Microplastics in the marine environment: A review of the methods used for identification and quantification. Environmental Science & Technology. 2012;**46**:3060-3075. DOI: 10.1021/es2031505

[17] Hann S, Sherrington C, Jamieson O, Hickman M, Kershaw P, Bapasola A, et al. Investigating Options for Reducing Releases in The Aquatic Environment of Microplastics Emitted By (but not intentionally added in) Products. 2017. Available from: http://www.eumicroplastics.com/eumpwp/wpcontent/uploads/investigating-options-eunomia-draftreport-v4-main-report-public.pdf [Accessed: November 4, 2017]

[18] Löder MGJ, Gerdts G. Methodology used for the detection and identification of microplastics—A critical appraisal. In: Bergmann M, Gutow L, Klages M, editors. Marine Anthropogenic Litter. Cham: Springer International Publishing; 2015. pp. 201-227. DOI: 10.1007/978-3-319-16510-3_8

[19] UNDP. Beyond Scarcity: Power, Poverty and the Global Water Crisis. New York, NY: UNDP; 2006

[20] Lattemann S, Höpner T. Environmental impact and impact assessment of seawater desalination. Desalination. 2008;**220**:1-15. DOI: 10.1016/j.desal.2007.03.009

[21] Maupin MM, Kenny JF, Huston SS, Lovelace JK, Barber NL, Linsey KS. Estimated use of water in the United States in 2010. In: Geological Survey Circular 1405. 2010. DOI: 10.3133/cir1405. 56 p

[22] Sha Z, editor. 9th International Symposium on Salt, held in Beijing International Convention Center; from September 4-7; Beijing, China. 2009

[23] Iñiguez ME, Conesa JA, Fullana A. Microplastics in Spanish table salt. Scientific Reports. 2017;**7**:8620. DOI: 10.1038/s41598-017-09128-x

[24] Yang D, Shi H, Li L, Li J, Jabeen K, Kolandhasamy P. Microplastic pollution in table salts from China. Environmental Science & Technology. 2015;**49**:13622-13627. DOI: 10.1021/acs.est.5b03163

[25] Hoepner T, Lattemann S. Chemical impacts from seawater desalination plants—A case study of the northern Red Sea. Desalination. 2003;**152**:133-140. DOI: 10.1016/S0011-9164(02)01056-1

[26] Bagher M. Desalination Oppurtunities in Iran, World Water-Tech Innovation Summit; 21 February 2018; London

[27] Bertling R, Bertling J, Hiebel M. Fraunhofer UMSICHT nimmt Stellung. Thema: Mikroplastik: Fraunhofer UMSICHT; 2017

[28] Ziccardi LM, Edgington A, Hentz K, Kulacki KJ, Kane Driscoll S. Microplastics as vectors for bioaccumulation of hydrophobic organic chemicals in the marine environment: A state-of-the-science review. Environmental Toxicology and Chemistry. 2016;**35**:1667-1676. DOI: 10.1002/etc.3461

[29] Lithner D, Larsson A, Dave G. Environmental and health hazard ranking and assessment of plastic polymers based on chemical composition. Science of the Total Environment. 2011;**409**:3309-3324. DOI: 10.1016/j.scitotenv.2011.04.038

[30] Lusher A, Hollman PCH, Mendoza-Hill J. Microplastics in Fisheries and Aquaculture: Status of Knowledge on their Occurrence and Implications for Aquatic Organisms and Food Safety. Rome: Food and Agriculture Organization of the United Nations; 2017

[31] Rafiee M, Dargahi L, Eslami A, Beirami E, Jahangiri-Rad M, Sabour S, et al. Neurobehavioral assessment of rats exposed to pristine polystyrene nanoplastics upon oral exposure. Chemosphere. 2018;**193**:745-753. DOI: 10.1016/j.chemosphere.2017.11.076

[32] Herbort AF, Schuhen K. A concept for the removal of microplastics from the marine environment with innovative host-guest relationships. Environmental Science and Pollution Research. 2016;**24**(12):1-5. DOI: 10.1007/s11356-016-7216-x

[33] Herbort AF, Sturm MT, Schuhen K. A new approach for the agglomeration and subsequent removal of polyethylene, polypropylene, and mixtures of both from freshwater systems—A case study. Environmental Science and Pollution Research International. 2018;**25**(15):15226-15234. DOI: 10.1007/s11356-018-1981-7

[34] Schuhen K. Hybridkiesel-säurematerial, Insbesondere Zur Fixierung Anthropogener Verunreinigungen Aus Einem Aquatischen Umfeld; April 17, 2015

[35] Herbort AF, Sturm MT, Fiedler S, Abkai G, Schuhen K. Alkoxy-silyl induced agglomeration: A new approach for the sustainable removal of microplastic from aquatic systems. Journal of Polymers and the Environment. 2018;**62**:1596. DOI: 10.1007/s10924-018-1287-3

[36] Wasser 3.0 www.wasserdreinull.de. Wasser 3.0: Mikroplastikentfernung. 2018. Available from: https://www.youtube.com/watch?v=svwN4pamXAg [Accessed: 24 May, 2018]

[37] Auner N, Weis J. Organosilicon Chemistry: From Molecules to Materials. Weinheim, New York: VCH; 1994

[38] Moreau JJE, Vellutini L, Man C, Wong M, Bied C, Bantignies J-L, et al. Self-organized hybrid silica with long-range ordered lamellar structure. Journal of the American Chemical Society. 2001;**123**:7957-7958. DOI: 10.1021/ja016053d

[39] Corriu R, Leclercq D. Neue Entwicklungen der Molekülchemie für sol-gel-prozesse. Angewandte Chemie. 1996;**108**:1524-1540. DOI: 10.1002/ange.19961081305

[40] Bernards TNM, van Bommel MJ, Boonstra AH. Hydrolysis-condensation processes of the tetra-alkoxysilanes TPOS, TEOS and TMOS in some alcoholic solvents. Journal of Non-Crystalline Solids. 1991;**134**:1-13. DOI: 10.1016/0022-3093(91)90005-Q

[41] Hurkes N, Ehmann HMA, List M, Spirk S, Bussiek M, Belaj F, et al. Silanol-based surfactants: Synthetic access and properties of an innovative class of environmentally benign detergents. Chemistry. 2014;**20**:9330-9335. DOI: 10.1002/chem.201402857

[42] Herbort AF, Schuhen K. Patent Pending; Notifying number 102018203185.1, Submission date 02.03.2018

[43] Fent K. Einfluss von umweltfaktoren: Temperatur, salinität und UV-strahlung. In: Fent K, editor. Ökotoxikologie. Stuttgart: Georg Thieme Verlag; 2013. DOI: 10.1055/b-0034-98977

[44] Rahman IA, Vejayakumaran P, Sipaut CS, Ismail J, Abu Bakar M,

Adnan R, et al. Effect of anion electrolytes on the formation of silica nanoparticles via the sol–gel process. Ceramics International. 2006;**32**:691-699. DOI: 10.1016/j.ceramint.2005.05.004

[45] Cook D. Lotus Illustrated Dictionary of Chemistry. New Delhi: Lotus Press; 2004

[46] Schwarzenbach RP, Gschwend PM, Imboden DM. Environmental Organic Chemistry. 2nd ed. Hoboken, N.J: Wiley; 2005

[47] Hench LL, West JK. The sol-gel process. Chemical Reviews. 1990;**90**: 33-72. DOI: 10.1021/cr00099a003

[48] Sturm MT, Wilde A, Kluczka S, Schuhen K. Detektion von Mikroplastik im (Ab-)Wasser – Die Suche nach der Nadel im Heuhaufen? Analytik-News – Das Online-Labormagazin. 2018;**5**:1-6

[49] Herbort AF, Sturm MT, Hiller C, Schuhen K. Nano- und Mikroplastik – Braucht es eine komplizierte Einzelstoffdetektion bei der Gewässeranalytik? Umdenken mit dem Wasser 3.0 – PEI?! GIT Labor-Fachzeitschrift. 2017;**3**:32-35

Chapter 3

Elemental Analyzer/Isotope Ratio Mass Spectrometry (EA/IRMS) as a Tool to Characterize Plastic Polymers in a Marine Environment

Daniela Berto, Federico Rampazzo, Claudia Gion, Seta Noventa, Malgorzata Formalewicz, Francesca Ronchi, Umberto Traldi and Giordano Giorgi

Abstract

In the last 60 years, plastic has become a widely used material due to its versatility and wide range of applications. This characteristic, together with its persistence, makes plastic waste a growing environmental problem, particularly in the marine ecosystems. The production of plant-derived biodegradable plastic polymers is assuming increasing importance. Here, we report the results of a first preliminary characterization of carbon stable isotopes ($\delta^{13}C$) of different plastic polymers (petroleum- and plant-derived) and a first experimental study aimed to determine carbon isotopic shift due to polymer degradation in an aquatic environment. The results showed that the $\delta^{13}C$ values determined in different packaging for food uses reflect the plant origin for "BIO" materials and the petroleum-derived source for plastic materials. Considering degradation, $\delta^{13}C$ values of both bio bags and HDPE bags showed a gradual decrease toward less negative values when kept immersed in seawater, recording a $\delta^{13}C$ variation ($\Delta\delta^{13}C$) of 1.15 and 1.78‰, respectively. With respect to other analytical methods, the characterization of the plastic polymer composition by isotope ratio mass spectrometry is advantageous due to low cost and rapidity of analysis, small amount of sample required, high sensitivity, and the possibility of analyzing colored samples.

Keywords: carbon isotopes, plastic polymers, EA/IRMS, plastic degradation, plastic pollution

1. Introduction

1.1 Plastic debris in the ocean: a global environmental issue of the twenty-first century

Since 1950, the production and use of plastics has been constantly increased reaching a global production of 280 million tons in 2016 (i.e., as thermoplastics and polyurethanes), with China as the major producer (29%) [1].

IntechOpen

Plastics represent a group counting hundreds of different materials derived from fossil sources (e.g., oil and gas) among which the most produced are polypropylene (PP), high- and low-density polyethylene (HDPE, LDPE), polyvinyl chloride (PVC), polyurethane (PUR), polyethylene terephthalate (PET), and polystyrene (PS). Due to their high versatility, durability, low weight, and low cost, plastic materials find applications in almost any market sector, but primarily in packaging (39.9%) and building industries (19.7%) [1].

In recent years, the growing evidence about the massive presence of plastic litter in the ocean, its pressure on the marine environment and wildlife, and its impact on marine-related human activities (such as fishery, shipping, and tourism) has raised lot of attention in the scientific, regulatory, and civil communities (**Figure 1**).

Oceanographic surveys have recorded the presence of plastics in any geographical regions, including remote polar areas, and at any depth, from the sea surface to the seafloor of the oceans (**Figures 2–4**).

The amount of plastic debris in the sea is still unknown due to the large variability of its distribution as regards both spatial and temporal scale, which prevents accurate estimates. However, modeling studies have recently approximated that 5–13 million tons of plastics (i.e., equivalent to 1.5–4% of global plastic production) end up in the oceans every year [2].

The slow degradation rates of plastics under environmental conditions provide additional complexity to this global issue, by contributing to their accumulation in all terrestrial and aquatic environments. It has been estimated that, once in the ocean, the majority of manufactured polymers persist for decades and probably for centuries due to their low degradability (**Figure 5**) [3, 4].

In both terrestrial and marine environments, degradation of petroleum-derived plastics occurs through abiotic and biotic processes (i.e., UV degradation, hydrolysis, and decomposition by microorganisms), leading to their fragmentation into increasingly smaller pieces. Thus, plastic particles dispersed in the environment are commonly divided into three main classes based on their size: macro: >25 mm,

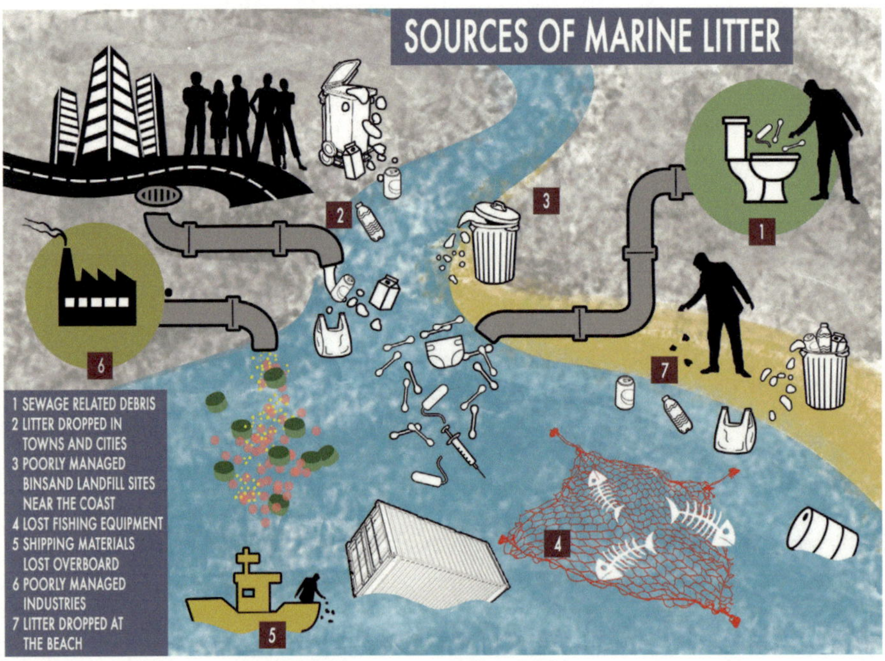

Figure 1.
Pathways through which litter reaches the sea (illustration by Davide Zanella).

meso: 5–25 mm, and microplastics: <5 mm. The smaller-size class, which includes both primary microplastics (i.e., particles produced as such, e.g., plastic pellets, exfoliating cosmetics, or synthetic clothing fibers) and secondary particles (i.e., particles derived from the breakdown of larger plastic debris), is likely to be the most abundant in the ocean today [5].

Figure 2.
Marine litter on the beach (photo by Tomaso Fortibuoni).

Figure 3.
Seabird nesting on plastic nets (public domain).

Figure 4.
Tangle of fishing nets on a beach (photo by Francesca Ronchi).

Figure 5.
Estimated decomposition times of different types of garbage dispersed in the marine environment (illustration by Davide Zanella).

The concern about the heavy contamination of the marine environment by plastics is related to the potential of plastic debris to cause harm to the inhabiting organisms via different mechanisms. Among the most alarming issues, there is an uptake and a bioaccumulation of plastic debris by marine organisms at almost all levels of the food web and the consequent trophic transfer. Recent studies have reported that micro and nanoplastics can easily be taken up and ingested by marine organisms (i.e., zooplankton, worms, bivalves, crustaceans, demersal and pelagic fishes, seabirds, reptiles, and mammals), resulting in a significant impact on the aquatic wildlife and possibly on human health via seafood consumption [6]. Furthermore, due to the large surface to volume ratio, microplastic fragments can potentially adsorb many kinds of common marine contaminants on their surface, in particular hydrophobic organic substances such as polychlorinated biphenyls, polyaromatic hydrocarbons, and organochlorine pesticides [7, 8]. This can promote their transport in the environment and induce toxic effects following ingestion and desorption (e.g., endocrine disruption, mutation, and cancer). Moreover, another source of concern is the possible release of additives commonly present in plastic formulations (i.e., bisphenol A, phthalates, and flame retardants) [8, 9], and although the leaching rates of these common additives in seawater are poorly known, their potential for toxicity is considered to be very high.

Several actions have currently been undertaking at national and international levels to tackle the contamination of marine environments by plastics. Their main aim is to achieve a general reduction of plastic use (in particular packaging and disposable items), recycling of plastic items at the end of their lifetime, and replacement of the use of plastics with more sustainable materials and biopolymers (e.g., plant-derived polymers [10]), which are more prone to degradation by microorganisms and show a shorter persistence once dispersed in the environment.

1.2 Experimental approaches to assess plastic debris in environmental samples

With the growing evidence of the severe impact caused by plastics on the wildlife, the assessment of the presence, behavior, and fate of plastics in the marine environments has become a fundamental research issue, highly advocated to the scope of putting in place more effective policies. However, especially for the smallest particles (i.e., microplastics), their efficient identification to the scope of assessing the plastic load in the environmental compartments (e.g., seawater, sediments, and biota) is a serious challenge for scientists. Many analytical techniques have been used to identify plastic debris in environmental samples, as largely reviewed in the literature [11, 12]. Among the most used approaches, there are electron scanning microscopy coupled with energy dispersive X-ray spectroscopy (SEM-EDS, ESEM-EDS), Raman spectroscopy, Fourier transform infrared spectroscopy (FT-IR) [13], and thermal analysis (pyro-GC/MS). Other analytical methods used to identify plastic materials are near infrared spectroscopy (NIRS), differential scanning calorimetry [14], and UV-VIS spectroscopy [15, 16].

Stable isotope analysis, which is an analytical technique that measures the relative abundance of stable isotopes yielding an isotope ratio that can be used as a research tool, is finding application in a growing number of different research fields and practical case studies. For instance, it is widely used to trace the origin of organic matter in various environments [17, 18], to track fraud in the food industries [19] and to identify microtraces of drugs, flammable liquids, and explosives in forensic cases [20]. This technique has been only rarely applied to assess the presence of microplastics in environmental samples [21]. Its potential for detecting plastic debris in environmental samples relies on different isotopic signatures of carbon in (i) petroleum-derived materials, (ii) C4 plants used in the synthesis of bioplastics, and (iii) marine samples' matrices (e.g., particulate organic matter, plankton, tissues of marine organisms, algae, and marine plants).

1.3 Stable isotope analysis: principles of the method

The term isotopes (from the Greek iso, same and topos, place) identifies atoms of the same chemical element, that is, the same place in the periodic table of the elements, that has the same atomic number but different atomic mass number. In other words, isotopes are atoms having the same number of protons and electrons (equal chemical properties) and a different number of neutrons (different physical properties). Each element has known isotopic forms, and in total, there are 275 isotopes of the 81 stable elements, in addition to over 800 radioactive isotopes (**Figure 6**).

Isotopes of a single element possess almost identical properties. They are commonly classified as natural or artificial, stable, or unstable. The quantification of the ratio between two isotopes allows to determine if two chemically similar environmental samples have different origins, related to the difference of the original sources. The isotopic distribution characterizing the sources may be influenced by phenomena of a different nature, which in turn may cause significant variations in the final products.

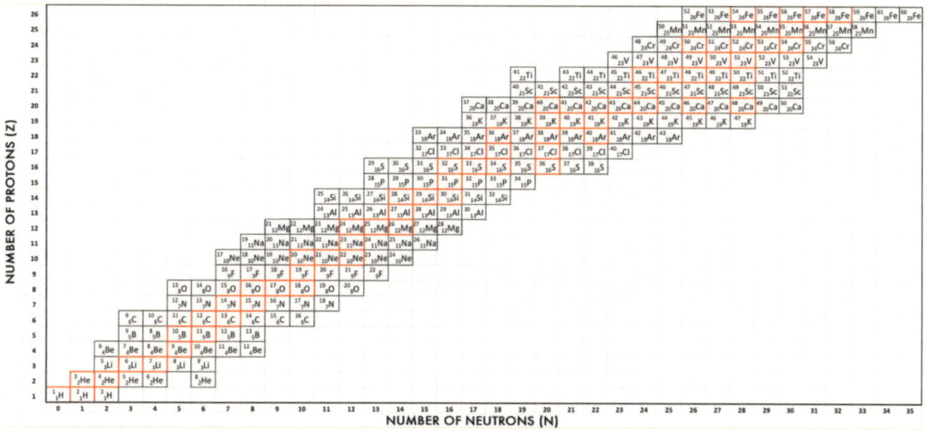

Figure 6.
Stable isotopes have a proton/neutron ratio lower than 1.5.

Depending on the chemical element, variations in the relative mass abundance of its isotopes can be detected through the analysis of stable isotopes. Technological advances in isotope analysis have led to the development of scientific instruments able to measure very small variations in the abundance of stable isotopes with high precision and accuracy (mass spectrometry). Therefore, stable isotope analysis can be applied considering different elements, thus giving nowadays applications in different fields of science.

For a given chemically stable element, its isotopic composition in a sample (R) is equal to the ratio between the abundance of the heavy isotope with respect to the light one (e.g., $^{13}C/^{12}C$), and it is expressed as deviation, in parts per thousand, from an international reference standard material (δ‰), according to the equation (Eq. (1)) given below:

$$\delta(\%) = \left[(R_{sample} - R_{standard})/R_{standard}\right] \times 1000 \quad (1)$$

where R_{sample} is the mass ratio of the heavy isotope to the light isotope measured in a sample and $R_{standard}$ is the isotopic ratio defined for the standard. The standard reference material that is commonly used for carbon is Vienna Pee Dee Belemnite. Thus, positive δ values indicate that the heavy isotope is enriched in the sample compared to the standard, while negative δ values indicate that the heavy isotope is depleted in the sample.

The possibility of distinguishing two samples on the basis of their relative abundance of two isotopes bases on the phenomenon of isotopic fractionation, which can be enacted by a wide range of chemical (e.g., nitrification and ammonification), physical (e.g., evaporation and condensation), and biological (e.g., photosynthesis, assimilation, and excretion) processes. In fact, many natural (and anthropic) processes can alter the isotopic signature of a chemical element in a matrix by causing an imbalance of the isotope distribution that leads to a variation of its original isotopic signature [22]. Thus, as the extent of fractionation of many chemical elements have been proved to be sensitive to specific processes/variables, it can be used as a tool to investigate the involved process/variable itself. In general, two mechanisms of isotopic fractionation can be distinguished:

(1) Thermodynamic, that is, due to a difference in bonding energy of the isotopes in the compounds. This mechanism implies that:
    - heavy isotopes accumulate in oxidized products;
    - the isotopic fractionation is favored at low temperatures, since at high temperatures, the differences between the isotopes are attenuated;
    - the process is not relevant in the case of chemical reactions of gaseous substances and biological reactions.

(2) Kinetic, that is, due to different reaction rates of the isotopes. In this type of mechanism, it is generally observed that fast, irreversible, and unidirectional processes are favored. This is the case of most of the biological processes, where lighter isotopes are preferentially used, as their employment requests less energy. It also occurs when the products are easily removed from the reagents, in processes such as evaporation, diffusion, etc. These processes are featured by:
    - the preferential breaking of the bonds formed by light isotopes;
    - the preferential distribution of light isotopes in products and of the heavy ones in the reagents.

Given a chemical substance AB characterized by the presence of a certain isotopic distribution of element X, we can calculate the fractionation factor by dividing the ratio of the number of isotopes X in product A with the ratio of the number of isotopes X in product B (Eqs. (2) and (3)).

$$\alpha_{AB} = \frac{R_A}{R_B} = 1 + \left[\frac{(\delta_A - \delta_B)}{1000}\right] \tag{2}$$

where

$$R = \frac{X_h[\textit{atoms of the heavier isotope (rare)}]}{X_l[\textit{atoms of the lighter isotope(abundant)}]} \tag{3}$$

However, the fractionation factor (α) is normally replaced by the isotopic enrichment factor (ε), which is defined as $(\alpha - 1) \times 1000$.

1.4 Carbon isotope ratio as a tool in environmental studies

Carbon and nitrogen isotope analysis is used to investigate the trophic web and the matter flows among the main components of an ecosystem (e.g., organic matter, producers, primary and secondary consumers); it can be used to understand chemical and biological processes occurring at both ecosystem and organism levels. Stable isotope analysis can also be a useful tool for assessing the origin of water, atmospheric, and soil pollution.

The two main carbon reserves in nature are represented by organic and inorganic carbon, which are characterized by different isotopic fingerprints due to the different processes in which they are involved (**Figure 7**). The inorganic carbon (carbonate) is involved in the exchange equilibrium among (i) atmospheric carbon dioxide, (ii) dissolved bicarbonate, and (iii) solid carbonate. The exchange reactions among these three forms lead to an enrichment of the heavy isotope in the

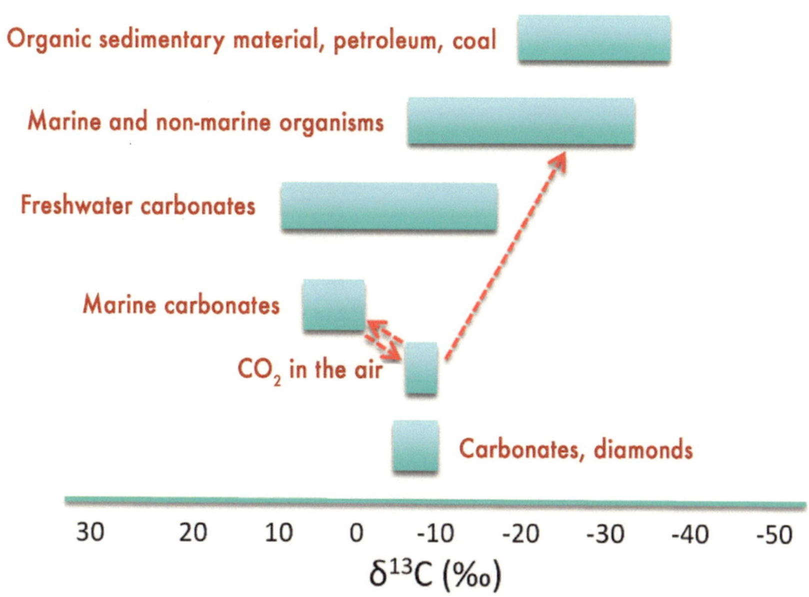

Figure 7.
Isotopic fingerprint of naturally occurring carbon.

solid carbonate form ($\delta^{13}C$ equal to 0‰). In contrast, the kinetic reactions which mainly involve the organic carbon (i.e., photosynthetic process) determine a concentration of the lightest isotope in the synthesized organic material ($\delta^{13}C$ equal to about −25‰) [17].

The fractionation of organic carbon is mainly linked to the specific photosynthetic pathway featuring each plant. The terrestrial plants, classified as C3 and C4, can follow two different photosynthetic pathways. Both types synthesize organic matter characterized by $\delta^{13}C$ values more negative than that of carbon dioxide (~−7‰), because during the photosynthesis, the produced organic substance accumulates the light isotope compared to the heavy one. The C3 plants, typical of temperate climates, produce the 3-phosphoglyceric acid, a compound with three carbon atoms (Calvin cycle) with an average value of $\delta^{13}C$ of about −26.5‰. The C4 plants generate oxaloacetate, a compound with four carbon atoms (Hatch-Slack cycle) characterized by a value of $\delta^{13}C$ around −12.5‰.

The chemical composition of animal tissues is related to the food sources they assimilate, and therefore, it reflects the isotopic composition of the diet [23, 24]. The enrichment between primary producers and consumers (herbivores) has been estimated to be approximately +5‰, whereas at the successive trophic levels, the enrichment is less marked (+1‰) [25]. Thus, the isotopic value detected in the tissues of an organism can be potentially used as an indicator of its trophic position. However, since the variation of the $\delta^{13}C$ values due to trophic passages is relatively modest, $\delta^{13}C$ is mainly used to trace the primary carbon source used [26].

Through the analysis of the stable carbon isotopes, it is also possible to differentiate terrestrial and marine trophic webs. The "marine" carbon derives from the dissolved inorganic carbon (dissolved bicarbonate) characterized by an isotopic value equal to about 0‰, while the "terrestrial" carbon derives from the atmospheric carbon dioxide which has a lower $\delta^{13}C$ value (approximately −7‰). This difference is maintained at every trophic level both in the marine and terrestrial trophic chain (**Figure 8**).

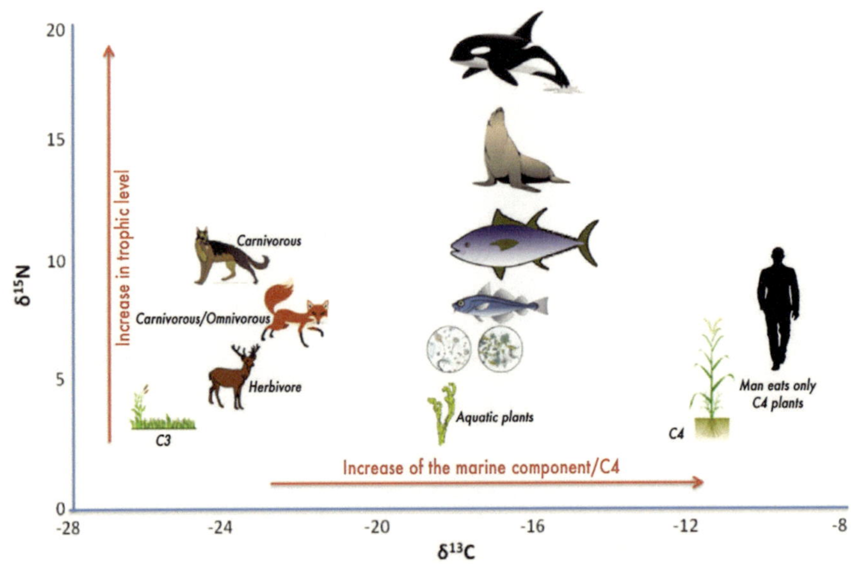

Figure 8.
Variations of δ¹³ carbon and δ¹⁵ nitrogen (‰) isotopes in different organisms of the terrestrial and marine food chain.

2. Preliminary study to characterize plastic polymers using elemental analyzer/isotope ratio mass spectrometry (EA/IRMS)

In 2016, Berto and collaborators carried out a preliminary study aimed at evaluating the potentials of stable isotope analysis to discriminate a wider range of plastic and bioplastic materials (including those highly used in packaging, such as shopping bags and plastic bottles for drinking water) (**Table 1**) to the scope of using this analytical technique for the identification of plastic debris in marine samples in future field surveys [27].

Furthermore, considering the lack of knowledge on possible changes in the carbon isotopic signature of plastics due to degradation processes in the marine environment, this study also investigated the variation of $\delta^{13}C$ values of petroleum- and plant-derived polymers of packaging materials subjected to biotic and abiotic degradation. The study was carried out by using an isotope ratio mass spectrometer Delta V Advantage (Thermo Fisher Scientific, Bremen, Germany) coupled with an elemental analyzer Flash 2000 (Thermo Fisher Scientific, Bremen, Germany). The accuracy of the isotopic data was evaluated by the analysis of the certified polyethylene foil (−31.8 ± 0.2‰, IAEA-CH-7, International Atomic Energy Agency, Austria). The analytical precision of measurements was 0.2‰ for C.

Sample type	Use
Acrylonitrile butadiene styrene (ABS)	Thermoplastic polymer; used for pipes, musical instruments, sport equipment, automobile industry, toys, and electronic component assemblies
Fluorocarbon	Polymers characterized by high melting temperature, resistance to sunlight degradation and chemical inertness; used for industrial coatings (electronics, automotive, and food industry)
High density polyethylene (HD PE)	Thermoplastic polymer; used for beverage bottles, food storage containers, plastic bags, fuel tanks, piping systems, electrical boxes, folding furniture, sport equipment, storage sheds, and wood-plastic composites; packaging material for food use
Low density polyethylene (LD PE)	Utilized as thin films used for coatings or plastic bag production, recipients made in injection molding (bottles, bags, and laboratory equipment), pipings, food containers, corrosion resistant surfaces, and computer components; packaging material for food use
Low density polyethylene (LD PE) recycled	As above
Chlorosulfonated polyethylene (HYPALON)	Is a synthetic rubber; used in folding kayaks and inflatable boats, sport equipment (snowshoes), and roof coatings
Polychloroprene (NEOPRENE)	This synthetic rubber is more degradation resistant than the natural one and is used for corrosion-resistant coatings and hoses, clothing and equipment for water sports, car seats or covers, computer or tablet holders, mouse pads, and gloves
Polyamide (PA)	Polymers used in carpets, textiles, sportswear, due to their strength and durability
Polyethylene terephthalate glycol-modified (PETG)	This strong and glass-like thermopolymer is employed in production of electronic devices, covers, food and medical applications, and credit cards; it is fully recyclable

Sample type	Use
Polypropylene (PP)	As it is chemically resistant to many acids, bases, and solvents, it is used for laboratory equipment, medical devices and textiles, but also for piping systems, plastic furniture, carpets, ropes, roofing waterproof membranes, insulation for electrical cables, bottles and storage boxes, and concrete additives
Polystyrene (PS)	In its solid form, it is employed for packaging, containers, bottles, disposable cutlery, dishes, and razors. Also widely used in foamed form in protective packaging
Polytetrafluoroethylene (PTFE)	This fluoropolymer is hydrophobic and nonreactive and displays an extremely low friction coefficient. It is used in industrial applications, wire production as insulators, carbon fiber and fiberglass composites production, a nonstick coating in cookware, and textiles (like Gore-Tex)
Polyurethane (PU)	Thermoplastic and thermosetting polymers are available; used for the production of cushions, mattresses (flexible foams), construction industry, wheels, tires, furniture, synthetic fibers, hard-plastic parts, shoes, and surface coatings
Polyvinyl chloride (PVC)	Production of door and window frames, siding, pipes, bottles, and credit cards (rigid form); also used in electrical cable insulation, inflatable products, leather imitation, and shower curtains (flexible form)
Polyvinylidene difluoride (PVDF)	Chemically, electrically and flame resistant resin, used for containers in chemical industry, pipes, and food containers. Also applied in electronics (transducers and insulators) and in medical and defense industries
Styrene acetonitrile (SAN)	Polymer similar to polystyrene; used in plastic optical fibers, packaging material, bottles, and food container, as well as in computer products and battery cases
Polysiloxanes (SILICONE)	Used in electrical and thermal insulation, sealants, medical applications, and cooking utensils
Polyethylene terephthalate (PET)	Used for packaging of food and beverages, textiles (known as polyester), and thermoforming applications
PET recycled	As above
LATEX/RUBBER	Plant-derived plastic polymer. The natural latex rubber is employed in the production of gloves, mattresses, balloons, and swim caps
Polylactide (PLA)	Is a thermoplastic, plant-derived polyester. Used as a packaging material, compost bags, disposable tableware, gardening (mulch film), and medicine (implants)
"BIO" bag	Biodegradable bags are used in food packaging and composting of food waste and to collect dog waste
Cellulose	C3 plant used in bioplastic production
Cotton	C3 plant used in bioplastic production.
Potato	C3 plant used in bioplastic production.
Rice	C3 plant used in bioplastic production
Tomato	C3 plant used in bioplastic production
Corn	C4 plant used in bioplastic production
Sugarcane	C4 plant used in bioplastic production

Table 1.
Plastic (petroleum and plant-derived polymers) and natural matrices analyzed in this study.

This chapter gives a review of the main insights obtained and critically discusses the potentials of the carbon isotope ratio analysis to study the behavior and fate of plastics in the aquatic environment.

2.1 $\delta^{13}C$ as a possible tool to investigate plastic polymers

$\delta^{13}C$ values recorded in this study for the most used petroleum-derived plastic polymers, plant-derived polymers, some commercial items made by petroleum- and natural-derived polymers, which are largely found in the marine litter worldwide (i.e., food packaging items), and natural matrices are reported in **Figure 9**.

Due to their high stability and durability [28], in the last decades, petroleum-derived plastic materials have largely replaced paper and other cellulose-based products with a continuously increasing trend. At the moment, a wide variety of petroleum-based synthetic polymers are produced worldwide (approximately a total of 335 million tons in 2016), and significant quantities of these polymers end up into natural ecosystems as waste products [1].

The $\delta^{13}C$ values of the majority of the analyzed petroleum-derived plastic polymers ranged over a wide interval, that is, between −33.97 and –25.41‰. Only a few polymers, such as PTFE, silicon, and ABS showed more negative $\delta^{13}C$ values (−40.70 ± 1.17, −39.37 ± 0.27, and − 35.17 ± 0.98‰, respectively), possibly due to fractionation processes during their synthesis.

With the exclusion of PTFE, ABS, and silicon, the recorded $\delta^{13}C$ range results are comparable to that reported for crude petroleum [29]. Petroleum is constituted by a complex mixture of organic substances, with a predominance of hydrocarbons, whose exact composition depends on the site of extraction. Petroleum usually shows negative values of $\delta^{13}C$, ranging between −34 and −18‰ depending on the specific extraction field. In fact, as reported by Stahl [29], petroleum could be originated from the lipid fraction of organic matter. In particular, the carbon

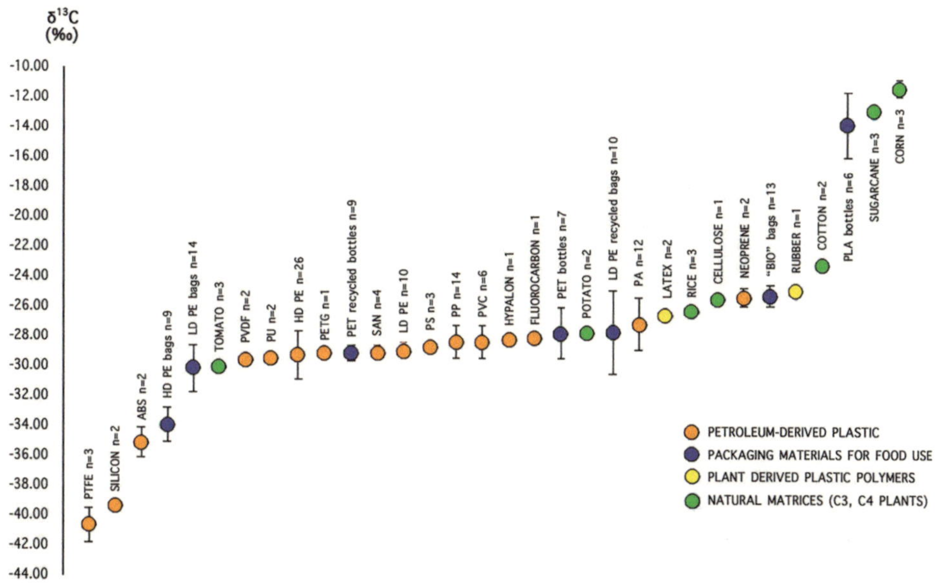

Figure 9.
$\delta^{13}C$ values determined in various petroleum- and plant-derived polymers, as well as in natural matrices analyzed in this study.

isotopic value of petroleum can vary in relation to the marine vs. terrestrial origin of the source, with an enrichment of ^{12}C with respect to ^{13}C in the marine environment compared to the terrestrial one [30].

Interestingly, different $\delta^{13}C$ values were recorded for some polymers as pure material and once in packaging commercial items. For instance, a significant ($p < 0.05$) more negative $\delta^{13}C$ value was determined in the HDPE shopping bag for food use with respect to the original HDPE polymer. This could be related to the addition of some organic additives (i.e., stabilizers) in the final materials used for food packaging. In fact, depending on the commercial use, plastic formulations can be enriched with monomeric ingredients to improve their processing, end-use performance, and appearance (e.g., colorants, photostabilizers, etc.). Among these possible additives, our preliminary data excluded colorants as the main cause of isotopic variation in the investigated samples. These results were confirmed by the lack of significant difference among polymers of different colors ($p > 0.001$). The independence of the $\delta^{13}C$ value from the plastic color could provide an important analytical advantage to the isotopic approach over some of the other analytical methods used for plastic characterization. In particular, the spectroscopic methods have been proved to be limited by the color of the plastic samples, because of the occurrence of interferences due to a decrease of the diffuse reflection intensity in dark color samples [31]. Further investigation and larger analytical data set are required in order to strength these results.

2.2 $\delta^{13}C$ as a possible tool to distinguish petroleum-derived plastics from plant-derived plastics

To reduce the impact of plastic debris in the environment, recyclable and more biodegradable polymers (i.e., plant-derived polymers) have been introduced increasingly into the market [10]. Plant-derived plastic polymers used for food packaging, such as bags and bottles for drinking water, showed a significant difference in isotopic values with respect to the petroleum-derived plastic products. In fact, petroleum-derived packaging materials for food use, such as shopping bags for fruits and vegetables (HD PE) and plastic bottles for drinking water (PET), were characterized by the $\delta^{13}C$ mean values of −33.97 ± 1.15 and − 27.84 ± 1.71‰, respectively, whereas plant-derived supermarket envelopes ("BIO" bags) and bottles (PLA, a biodegradable polyester derived from the fermentation of starch and condensation of lactic acid) recorded the $\delta^{13}C$ mean values of −25.30 ± 0.70 and −13.87 ± 2.18‰, respectively. As regards to the results obtained for "BIO" bags, values reflected those of C3 plants, while for PLA, the analyses highlighted $\delta^{13}C$ values similar to those of C4 plants, suggesting their specific origin.

This difference suggests that stable isotope analysis could be a useful method to discriminate between petroleum-and plant-derived plastic debris [21, 27]. The most used biopolymers are in fact produced starting from C3 (rice, potatoes, cotton, and cellulose) and C4 (corn and sugarcane) plants, species which differ for photosynthetic pathways and, consequently, for the carbon fingerprint. C3 plants recorded more negative $\delta^{13}C$ values (ranging from −30 to −25‰) than C4 plants (ranging from −13 to −11‰), in agreement with Suzuki et al. and authors therein [21]. Considering the isotopic signature of the "BIO" bags, a common and widespread biodegradable product used for many commercial purposes, $\delta^{13}C$ values are generally comparable with those reported for C3 plants. Regarding "recycled" polymers, LD PE recycled envelopes showed a $\delta^{13}C$ mean value of −27.75‰. The presence of a low quantity of other polymers as impurities or different recycle processes could explain the less negative average value with respect to the row LD PE (−30.19‰) given by an ^{13}C enrichment or depletion (fractionation).

2.3 $\delta^{13}C$ as a possible tool to study degradation processes in the marine environment

The influence of natural degradation processes on the fractionation of carbon in plastic materials under marine conditions, according to a preliminary field study carried out by Berto et al. [27] in Venice lagoon, is showed in **Figure 10**. Over a 60-day period and under variable conditions of temperature and salinity (i.e., 24–35°C and 7.8–8.1, respectively), the $\delta^{13}C$ values of both "BIO" bags and HD PE bags showed a gradual decrease toward less negative values, recording a $\delta^{13}C$ variation ($\Delta\delta^{13}C$) of 1.15 and 1.78‰, respectively. This shift could be reasonable due to physical, chemical, and/or biological degradation, even if the latter is a controversial matter.

The degradation of plastic polymers in the environment involves many factors (photodegradation, thermooxidation, hydrolysis, and biodegradation by microorganisms) [32], and it proceeds according to the rates highly dependent on the environmental conditions. For instance, several authors have reported that degradation processes and the rate of hydrolysis of most plastic polymers become insignificant in the ocean when the temperature and the concentration of oxygen are reduced [32, 33].

However, the physical/mechanical degradation occurring in the marine environment can alter the plastic polymers at the surface layer and favor the starting of microbial deterioration processes. By considering that, in many biochemical reactions, such as autotrophic fixation of CO_2 by plants [34] and microbial degradation processes, the lightest isotope (^{12}C) are preferentially used as a substrate over the heaviest isotopes, and the different isotopic values recorded by Berto et al. [27] for "pristine" and "aged" plastic materials sampled from the marine environment suggested the occurrence of degradation processes. Further studies are needed to evaluate the pathway and the time featuring this process.

In fact, some researchers are confident in thinking that biopolymer (such as cellulose in plants) plastics are not generally biodegradable. Bacteria and fungi coevolved with natural materials, while plastics have only been around for about 70 years. So microorganisms simply have not had much time to evolve the necessary biochemical tool kit to latch onto the plastic fibers, break them up into the constituent parts, and then use the resulting chemicals as a source of energy and carbon that they need to grow [35]. However, in 2016 a team of researchers from Kyoto Institute of Technology and Keio University, after collecting environmental samples

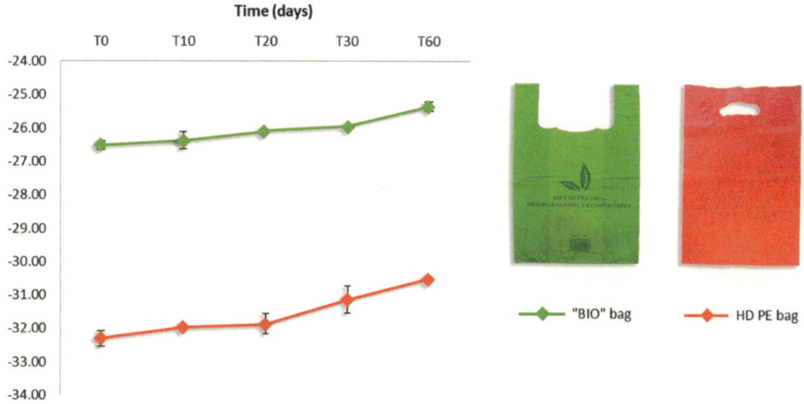

Figure 10.
Degradation of "BIO" bag and HD PE bag in lagoon environment, reflected by $\delta^{13}C$ variation in time.

containing PET debris, observed a novel bacterium (*Ideonella sakaiensis* 201-F6) which is able to use PET plastic for carbon growth. This bacterium produces two distinct enzymes hydrolizing PET plastics into terephthalic acid and ethylene glycol. This discovery has potential importance for the recycling process of PET [36].

A large number of tests (respirometric, loss of weight, tensile strength, spectroscopic) have been conducted to evaluate the extent of degradation of polymers, either alone or in blended forms, mainly under terrestrial environmental conditions.

It is worth noting that most recalcitrant polymers can be degraded to some extent in the appropriate environment at the right concentration. A screening program to study the ability of organisms and enzymes in degrading plastic polymers in a marine environment is required, considering the increasing importance of biodegradable plastics in the last few years.

Considering the new data presented in this study, it is possible to hypothesize the new paths for stable isotope research applied to the plastic polymers in the environment.

3. Conclusive remarks

In this chapter, we focused on plastic polymers, both petroleum- and plant-derived, commonly used in commercial packaging products for food use, giving preliminary overview of their $\delta^{13}C$ values. The low difference of $\delta^{13}C$ values among polymers suggested that the different chemical pathways used for their synthesis did not induce fractionation of carbon stable isotopes, yielding to $\delta^{13}C$ values meaningful of the row material (i.e., petroleum and terrestrial plants). Thus, this technique showed interesting perspective for its application in discriminating petroleum- and plant-derived polymers in marine samples.

Furthermore, the method showed to be unaffected by additional variables, such as color, and thus, it seems a valuable alternative to the spectroscopy methods for the characterization of plastic polymers in marine samples, which in contrast found the analytical limitation especially with dark colored plastic samples.

Finally, an important potential of the isotope mass spectrometry is its application to the study of the degradation processes (abiotic and biotic) of plastic waste released in the marine environment and the assessment of the degradation rates. In particular, this technique could be applied for analysis of suspended plastic debris, after filtration of both marine and fresh water samples collected along the water column. In this regard, however, further studies are needed to discriminate the isotopic values of suspended organic matter from those of plastic polymers, with major concern for micro and nanoplastics. Such possible application is of particular interest for the estimation of the fate of plastics in the marine environment and the evaluation of the effectiveness of the policies developed to reduce the environmental impact of marine litter.

Acknowledgements

The authors are grateful to Davide Zanella for graphical support and to Guido Giazzi and Luca Simonotti of Thermo Fischer Scientific for analytical support.

Conflict of interest

No potential conflict of interest was reported by the authors.

Author details

Daniela Berto[1]*, Federico Rampazzo[1], Claudia Gion[1], Seta Noventa[1], Malgorzata Formalewicz[1], Francesca Ronchi[1], Umberto Traldi[2] and Giordano Giorgi[3]

1 Italian National Institute for Environmental Protection and Research (ISPRA), Chioggia (VE), Italy

2 Thermo Fisher Scientific, Rodano (MI), Italy

3 Italian National Institute for Environmental Protection and Research (ISPRA), Rome, Italy

*Address all correspondence to: daniela.berto@isprambiente.it

References

[1] PlasticsEurope. Plastics—The Facts 2017: An Analysis of European Plastics Production, Demand and Waste Data; 2017. https://www.plasticseurope.org/application/files/5715/1717/4180/Plastics_the_facts_2017_FINAL_for_website_one_page.pdf

[2] Jambeck JR, Geyer R, Wilcox C, Siegler TR, Perryman M, Andrady A, et al. Plastic waste inputs from land into the ocean. Science. 2015;**347**(6223): 768-771. DOI: 10.1126/science.1260352

[3] Klein S, Dimzon IK, Eubeler J, Knepper TP. Analysis, occurrence, and degradation of microplastics in the aqueous environment. In: Wagner M, Lambert S, editors. Freshwater Microplastics: Emerging Environmental Contaminants? Cham: Springer International Publishing; 2018. pp. 51-67

[4] UNEP_YearBook. Plastic Debris in the Ocean. In: Emerging Issues in Our Global Environment [Internet]. 2011. Nairobi: United Nations Environment Program

[5] Law KL, Thompson RC. Microplastics in the seas. Science. 2014;**345**(6193):144-145. DOI: 10.1126/science.1254065

[6] Gall SC, Thompson RC. The impact of debris on marine life. Marine Pollution Bulletin. 2015;**92**(1):170-179. DOI: 10.1016/j.marpolbul.2014.12.041

[7] Rios LM, Moore C, Jones PR. Persistent organic pollutants carried by synthetic polymers in the ocean environment. Marine Pollution Bulletin. 2007;**54**(8):1230-1237. DOI: 10.1016/j.marpolbul.2007.03.022

[8] Bowmer T, Kershaw P. Proceedings of the GESAMP International Workshop on Plastic Particles as a Vector in Transporting Persistent, Bio-Accumulating and Toxic Substances in the Oceans; GESAMP; 2010. p. 68

[9] Zhang H, Zhou Q, Xie Z, Zhou Y, Tu C, Fu C, et al. Occurrences of organophosphorus esters and phthalates in the microplastics from the coastal beaches in north China. Science of The Total Environment. 2018;**616-617**:1505-1512. DOI: 10.1016/j.scitotenv.2017.10.163

[10] Auras R, Harte B, Selke S. An overview of polylactides as packaging materials. Macromolecular Bioscience. 2004;**4**(9):835-864. DOI: 10.1002/mabi.200400043

[11] Rocha-Santos T, Duarte AC. A critical overview of the analytical approaches to the occurrence, the fate and the behavior of microplastics in the environment. Trends in Analytical Chemistry. 2015;**65**:47-53. DOI: 10.1016/j.trac.2014.10.011

[12] Shim WJ, Hong SH, Eo SE. Identification methods in microplastic analysis: A review. Analytical Methods. 2017;**9**(9):1384-1391. DOI: 10.1039/C6AY02558G

[13] Reddy MS, Shaik B, Adimurthy S, Ramachandraiah G. Description of the small plastics fragments in marine sediments along the Alang-Sosiya ship-breaking yard, India. Estuarine, Coastal and Shelf Science. 2006;**68**(3):656-660. DOI: 10.1016/j.ecss.2006.03.018

[14] Wang JJ, Hou DX, Wu BL, Ye SL, Yang H. An effective classification procedure for plastic food packaging based on NIRS. Applied Mechanics and Materials. 2011;**88-89**:399-403. DOI: 10.4028/www.scientific.net/AMM.88-89.399

[15] Al Momani FA, Örmeci B. Measurement of polyacrylamide polymers in water and wastewater using an in-line UV–vis spectrophotometer.

Journal of Environmental Chemical Engineering. 2014;**2**(2):765-772. DOI: 10.1016/j.jece.2014.02.015

[16] de Lucia GA, Caliani I, Marra S, Camedda A, Coppa S, Alcaro L, et al. Amount and distribution of neustonic micro-plastic off the western Sardinian coast (Central-Western Mediterranean Sea). Marine Environmental Research. 2014;**100**:10-16. DOI: 10.1016/j.marenvres.2014.03.017

[17] Michener R, Lajtha K. Stable Isotopes in Ecology and Environmental Science. 2nd ed. Malden, MA, USA: Wiley-Blackwell Publishing Ltd.; 2007

[18] Berto D, Rampazzo F, Noventa S, Cacciatore F, Gabellini M, Aubry FB, et al. Stable carbon and nitrogen isotope ratios as tools to evaluate the nature of particulate organic matter in the Venice lagoon. Estuarine, Coastal and Shelf Science. 2013;**135**:66-76. DOI: 10.1016/j.ecss.2013.06.021

[19] Camin F, Boner M, Bontempo L, Fauhl-Hassek C, Kelly SD, Riedl J, et al. Stable isotope techniques for verifying the declared geographical origin of food in legal cases. Trends in Food Science and Technology. 2017;**61**:176-187. DOI: 10.1016/j.tifs.2016.12.007

[20] Meier-Augenstein W. Forensic Isotope Analysis in McGraw-Hill Yearbook of Science & Technology; McGrow-Hill Editor. New York, USA; 2014. p. 120-124

[21] Suzuki Y, Akamatsu F, Nakashita R, Korenaga T. A novel method to discriminate between plant- and petroleum-derived plastics by stable carbon isotope analysis. Chemistry Letters. 2010;**39**(9):998-999. DOI: 10.1246/cl.2010.998

[22] Brand WA, Coplen Tyler B, Vogl J, Rosner M, Prohaska T. Assessment of international reference materials for isotope-ratio analysis (IUPAC Technical Report). Pure and Applied Chemistry. 2014;**86**(3):425-467

[23] DeNiro MJ, Epstein S. Influence of diet on the distribution of carbon isotopes in animals. Geochimica et Cosmochimica Acta. 1978;**42**(5):495-506. DOI: 10.1016/0016-7037(78)90199-0

[24] Vander Zanden MJ, Rasmussen JB. Variation in $\delta^{15}N$ and $\delta^{13}C$ trophic fractionation: Implications for aquatic food web studies. Limnology and Oceanography. 2001;**46**(8):2061-2066. DOI: 10.4319/lo.2001.46.8.2061

[25] Tykot RH. Stable isotopes and diet: You are what you eat. In: Proceedings-International School of Physics Enrico Fermi. Vol. 154. Ohmsha: IOS Press; 2004. pp. 433-444

[26] Layman CA, Araujo MS, Boucek R, Hammerschlag-Peyer CM, Harrison E, Jud ZR, et al. Applying stable isotopes to examine food-web structure: An overview of analytical tools. Biological Reviews of the Cambridge Philosophical Society. 2012;**87**(3):545-562. DOI: 10.1111/j.1469-185X.2011.00208.x

[27] Berto D, Rampazzo F, Gion C, Noventa S, Ronchi F, Traldi U, et al. Preliminary study to characterize plastic polymers using elemental analyser/isotope ratio mass spectrometry (EA/IRMS). Chemosphere. 2017;**176**:47-56. DOI: 10.1016/j.chemosphere.2017.02.090

[28] Rivard C, Moens L, Roberts K, Brigham J, Kelley S. Starch esters as biodegradable plastics: Effects of ester group chain length and degree of substitution on anaerobic biodegradation. Enzyme and Microbial Technology. 1995;**17**(9):848-852. DOI: 10.1016/0141-0229(94)00120-G

[29] Stahl WJ. Carbon Isotopes in Petroleum Geochemistry. Berlin, Heidelberg: Springer; 1979

[30] Faure J, Mensing T. Isotopes Principles and Applications. 3rd ed. Hoboken, New Jersey: John Wiley & Sons Inc.; 2005. pp. 753-802

[31] Masoumi H, Mohsen Safavi S, Khani Z. Identification and classification of plastic resins using near infrared reflectance spectroscopy. International Journal of Mechanical and Industrial Engineering. 2012;**6**:213-220

[32] Andrady AL. Microplastics in the marine environment. Marine Pollution Bulletin. 2011;**62**(8):1596-1605. DOI: 10.1016/j.marpolbul.2011.05.030

[33] Raquez JM, Bourgeois A, Jacobs H, Degée P, Alexandre M, Dubois P. Oxidative degradations of oxodegradable LDPE enhanced with thermoplastic pea starch: Thermo-mechanical properties, morphology, and UV-ageing studies. Journal of Applied Polymer Science. 2011;**122**(1):489-496. DOI: 10.1002/app.34190

[34] O'Leary MH. Carbon isotope fractionation in plants. Phytochemistry. 1981;**20**(4):553-567. DOI: 10.1016/0031-9422(81)85134-5

[35] Lorch M. New plastic-munching bacteria could fuel a recycling revolution. GMT the Conversation. 2016: https://theconversation.com/new-plastic-munching-bacteria-could-fuel-a-recycling-revolution-55961

[36] Yoshida S, Hiraga K, Takehana T, Taniguchi I, Yamaji H, Maeda Y, et al. A bacterium that degrades and assimilates poly(ethylene terephthalate). Science. 2016;**351**(6278):1196-1199. DOI: 10.1126/science.aad6359

Chapter 4

Study of the Technical Feasibility of the Use of Polypropylene Residue in Composites for Automotive Industry

Denis R. Dias, Maria José O. C. Guimarães, Christine R. Nascimento, Celio A. Costa, Giovanio L. de Oliveira, Mônica C. de Andrade, Ana Maria F. de Sousa, Ana Lúcia N. da Silva and Elen B. A. Vasques Pacheco

Abstract

Polypropylene (PP) is widely used in short-term use artifacts, rapidly discarded and should partially replace neat PP. In addition, it is one of the polymers most used in the automobile industry. This study shows the technical feasibility of partially substituting neat PP for a post-consumer counterpart (PP_r), as well as adding ground glass (GP), used as filler in the polymer matrix. Mechanical and thermal properties of the recycled blends (PP/PP_r) and composites ($PP/PP_r/GP$) were evaluated. The results demonstrated that the blend with the highest PP_r content obtained a statistically significant decline in elastic modulus, but adding 5 wt% of GP to this blend increased this property, achieving a similar value in relation to neat PP. The composite developed may be a promising tailor-made product with properties resembling those of the virgin plastic. Thus, the automotive industry seems to be a good option for the use of PP_r and GP composites and blends, without increasing product requirements.

Keywords: recycling, polypropylene, automotive, industry, PP, glass powder, composite

1. Introduction

The circular economy promotes sustainability by combining the hierarchy of reduction, reuse and recycling, considering economic and environmental issues [1]. The use of post-consumer materials to manufacture new ones for a new production cycle to minimize the waste of natural resources is the goal of the circular economy [2–6]. The same can be said of plastic, an important class of materials that meets many of society's needs.

Plastics are present in our daily life under different forms and applications, such as office supplies, toys, footwear, civil construction, electrical and electronic components, aerospace, food, the medical and textile industries, packaging, paint and varnish, as well as the automotive industry, among others. This array of applications is due to their desirable properties in a range of sectors [7, 8].

However, not all plastics return to the production cycle after use. In 2015, only 9% of plastics produced worldwide were recycled, 12% incinerated and the rest buried in landfills [9]. Recycling causes less environmental impact, as reported by Bernardo et al. [10], who assessed recycling in terms of global warming and total energy use, concluding that plastic materials generally display environmental and economic advantages over conventional materials throughout their life cycle, from raw material extraction to synthesis, transformation, transport, use, recovery and destination. Duval and Maclean [11] also found a decline in greenhouse gas emissions and energy required during recycling [12]. Mechanical recycling involves the addition of virgin or recycled material to maintain properties [13].

Plastics can be recycled in different ways, including mechanically, chemically and energetically. Chemical recycling involves physical processes, such as remolding [14–16], and the final product is a monomer or oligomer that can be used in the synthesis of other products. In energy recycling, the energy released from the burning of waste material is reused [14].

Closed-loop recycling occurs when the recycled material replaces the virgin material in the same production cycle as the original product [17, 18]. Open-loop recycling is when the recycled product is used in a different production cycle, that is, the product to be recycled is used to manufacture a product different from the original [17, 19, 20].

A crucial point to stimulate recycling is the search for a different market for the recycled material and more environmentally sustainable processes. Traditionally, recycled products compete with virgin material, which may hinder their market entry. Scientific studies that focus on recycling should also seek to obtain more economically feasible and technically useful recycled products.

The most widely used and manufactured plastics are high-density polyethylene (HDPE), polypropylene (PP), low-density polyethylene (LDPE), poly(ethylene terephthalate) (PET), poly(vinyl chloride) (PVC) and polystyrene (PS) [9]. In the automotive industry, PP is used to manufacture the following items: car trunk lids; battery trays and boxes; heater boxes; tool boxes; seat belt buckling boxes; rear view mirror boxes; electric junction boxes; hubcaps; carpets; battery guards (protection against short circuit); steering wheel covers; shock absorber covers; vacuum hoses; air hoses; consoles; bumpers; glove boxes, among several other uses [21–24].

To comply with the main technical demands of automobile manufacturers, PP compounds must exhibit a suitable balance between stiffness and tenacity, with good thermal resistance, as well as fewer imported raw materials, thereby achieving more competitive prices. In addition to these properties, PP shows good processability [24–26].

An important supplier of materials to the automotive industry is the industrial sector responsible for manufacturing laminated and tempered glass used in motor vehicle windows (laminated glass for windshields and tempered for the other windows). However, in the tempering and laminating processes an industrial residue consisting of glass powder is generated and disposed of in landfills, with no specific use for this material [27]. In addition to the origin of glass powder in laminating and tempering processes [28], the windows that are removed from automobiles are also discarded when they cannot be reused. In such cases, these parts can be collected and recovered, then submitted to separation processes (polymer protection film) and grinding. The glass powder produced can be incorporated into polymer materials, resulting in composites with different properties [29].

Incorporating mineral loads into PP has been the object of studies on the production of materials with different properties [30, 31]. Improving the properties of the final product depends on the type of load, particle size of the mineral load being used and degree of dispersion of these particles in the polymer matrix. The most widely used commercial mineral loads are talcum and calcium carbonate [32–34].

This study describes the addition of glass powder to a PP matrix in order to obtain reinforcement properties and compare them with those of conventional composites. The aim is to acquire different properties in the polypropylene composites and reuse a residue (in this case, glass powder). We also assessed the effect of adding recycled polypropylene on the final properties of composites in order to reuse both industrial (glass powder) and urban residue (PP recycled from packaging).

2. The use of polypropylene in a composite or mixture

Polymers have been increasingly used in a number of applications as a substitute for traditional materials such as metal and ceramic, as homopolymers; formulated with additives, in the form of mixtures and polymer composites; or simply for their different properties, such as lightness, low transformation cost, resistance to corrosion, optimal thermal and electric insulation and easy conformation into complex shapes [33].

In general, the mechanical properties of polymers are not suitable in a number of applications owing to their lower resistance compared to metals and ceramics. However, the thermoplastic industry is growing due to ecological issues, in addition to the promising potential of these materials as mixtures or a composite matrix [35].

Compound systems formed by the combination of two polymer materials (mixtures) or a polymer material and a load (composites) are of significant technological interest due to the cost–benefit ratio. In both cases, the material consists of a continuous (matrix) and disperse phase, whose properties depend on good interaction between them [35].

The properties of interest for the automotive industry can be modified with studies on improving the polymer matrix, load, and polymer-load interface, among others. The interface is a link between the surface of the load and the matrix, and since the matrix receives the reinforcement, there is close contact between them, and there may or may not be adhesion. For a same combination of materials, different adhesion mechanisms can occur, such as mechanical, chemical, and electrostatic adhesion and by interdiffusion. The degree of reinforcement or improvement in mechanical behavior depends on a strong matrix-particle interface bond [36, 37].

The stress–strain behavior of many reinforced polymers or plastics can be changed by adhesion promoters and interfacial coupling agents (such as maleic anhydride) that alter adhesion and the nature of the matrix-load interface [38].

Polypropylene (PP) is a recyclable thermoplastic, that is, it melts when heated and hardens again when cooled, in a reversible process. Moreover, PP is easily mixed, primarily with organic reinforcing loads such as natural or inorganic fibers, including calcium carbonate, clay and talcum, and is widely used in structural applications [39–41].

The use of modified PP, especially for applications in the automotive industry, requires a suitable balance between stiffness and tenacity. In this scenario, the process of incorporating elastomeric materials, as well as mineral loads such as talcum and calcium carbonate ($CaCO_3$), into the PP matrix has been widely used to achieve different properties [42, 43].

Nanofillers, such as silica and calcium carbonate nanoparticles, have been added to improve the final properties of the PP matrix [44, 45].

2.1 Use of glass as an additive to the composite

The use of glass in a polypropylene matrix has been extensively studied and employed its glass fiber form in materials in which mechanical properties such as tensile strength and resistance to impact are important [46, 47].

There are several groups of glass, including silica, oxynitride and phosphate, but the first is the most important raw material used in composites [48]. Short E-glass fibers, obtained from a mixture of Si, Al, B, Ca and Mg oxides, are normally used as reinforcement for thermoplastics due to their low cost when compared to aramid and carbon [49], in addition to better impact strength and stiffness [50].

The interfacial interaction of glass composites with a thermoplastic matrix is often very weak. Particularly with polyolefin polymers such as polypropylene, there is little or no chemical reaction between the glass and the matrix. The interest in polypropylene for applications as a matrix in composites has been growing and the adhesion of this nonpolar polymer to the glass surface, which is also nonpolar, is a daunting challenge [51, 52].

In addition to the use of glass fiber, there are also glass microsphere applications [53]; however, residual glass powder remains a poorly explored load as reinforcement.

2.2 Environmental justification for polypropylene and window glass, materials contained in automobiles

Initiatives to develop more sustainable technological innovations and ecologically responsible management programs have been driven by a growth in environmental awareness and increasingly rigid legislation. The accumulation of plastic waste caused by the increase in per capital consumption of thermoplastic resins has prompted enormous research and efforts to substitute traditional thermoplastics [54].

To improve the production process, it is necessary to diagnose the flowchart of the process and manage inputs (water, energy, raw materials, etc.) and outputs (products, residues, effluents, atmospheric emissions, etc.). In general, inputs are natural resources that often cause environmental impact, such as ecosystem destruction, atmospheric pollution, etc. Outputs are environmental liabilities created by activities and residual materials (solid, liquid or gas) that, if not suitably managed, may cause permanent environmental impacts [55].

A sustainable production process contains a circular flow, where outputs are reintegrated into the process, which reduces impacts and costs in the generation of inputs and the destination and treatment of outputs. Recycling is an example of this type of sustainable production strategy and is therefore an attempt to reuse the material, natural resources and entropy expenditures in the production of a solid residue, reintroducing it into a new production process, thereby transforming the output of a process into the input of the same or another process [55].

The automotive industry is attempting to transform the car into a more sustainable and efficient product, not only in terms of the environment, but also from the consumer's financial standpoint. As such, the automotive industry has been working within the so-called DFE (Design for the Environment), that is, designing for the environment and introducing environmental variables in all the production strategies of the factory, such as product design (automobiles and parts), the process (manufacture of parts and assembly) and associated technologies (material treatment, painting, etc.) [28].

It is important to underscore that all participants in the life cycle of a product have shared responsibility. Thus, manufacturers, importers, distributors, merchants, consumers and public cleaning concessionaires should promote the reuse of solid residues, transfer them to the production chain, reduce residue generation and encourage the development of products derived from recycled materials. The automotive industry, like all companies, is responsible for the entire process, from acquiring raw materials to discarding components, such as bumpers. Moreover, polypropylene is present in many automobile components.

3. Materials and methods

3.1 Materials

The grade of the Virgin polypropylene (PP) was H 605. The values for the PP properties presented in **Table 1** were provided by the Braskem Company.

The post-industrial polypropylene (PPr) was a washed and ground material supplied by the Poli Injet Company (Brazil), while the residue came from the packaging industry. The properties of PPr are: melt temperature (Tm) of 161°C, crystallinity degree (χc) of 32%, and melt flow index (MFI) of 4.83 g/10 min (230°C/2.16 Kg). The methodologies to evaluate these properties are described below.

The glass powder (GP) used in this study was supplied by the Massfix Company (Brazil) and is ground from windshield scraps. **Figure 1** shows the morphology of GP samples examined under a scanning electron microscope (SEM). The GP sample is composed of irregular-shaped particles with a broad size distribution.

Polypropylene modified with maleic-anhydride (PPMA), and Polybond 3200 with 1 wt% of maleic anhydride (MA) were supplied by the Chemtura Company (USA). The melt flow rate specified by the supplier is 115 g/10 min at 190°C under 2.16 Kg.

3.2 Blend and composite preparation

Table 2 shows the compositions of the blends and composites. GP, PP, PPr, and PPMA were dried in an oven at 60°C for 24 h before extrusion. Next, each composition was processed in a twin-screw extruder (TeckTril, L/D = 36, screw diameter = 20 mm) at a screw speed of 400 RPM and temperature profile of 90/120/150/160/185/200/220/240/260/260°C. The materials underwent injection molding to produce appropriate specimens for stress and impact strength tests, which were performed according to ASTM D638 and ASTM D256 standards, respectively. Injection molding was carried out in an Arburg 270 S injection machine, using a

Properties	ASTM standard	Property value
Density, g/cm^3	D 792	0.905
MFI (230°C, 2.16 Kg), g/10 min	D 1238	2.1
Flexural modulus (1%), MPa	D 790	1600
Yield stress, MPa	D 638	37
Rockwell hardness (R)	D 785	101
Impact strength (Izod, 23°C)	D 256	45

BRASKEM company data sheet.

Table 1.
Polypropylene H605 properties.

Figure 1.
SEM micrographs of glass powder (GP) particles.

Component (%wt)	PP	PPr1	PPr3	PPr1/ GP	PPr3/ GP	PPr1/ GP/ PPMA	PPr3/ GP/ PPMA
PP	100	90	70	85	65	75	55
PPr	—	10	30	10	30	10	30
GP	—	—	—	5	5	5	5
PPMA	—	—	—	—	—	10	10

Table 2.
PP, PPr, glass powder and PPMA blends and (or) composites composition.

temperature profile of 210/215/220/230/230°C. The injection and molding pressures were 1000 and 180 Bar, respectively. The mold temperature was 30°C, with a cooling time of 30 seconds.

3.3 Material characterizations

Tensile properties were measured using a universal testing machine (EMIC, DL3000) based on ASTM D-638. Izod pendulum impact resistance was determined using a CEAST Resil Impactor tester based on ASTM D256. The data related to all the mechanical properties were based on the average of eight tested specimens.

Melting temperature (Tm), melt enthalpy (ΔHm) and crystallinity degree (χc) of the materials were determined using a Differential Scanning Calorimeter (STA 6000, Perkin Elmer) during the second heating scan. Samples weighing between 25 and 30 mg were heated from room temperature to 300°C at a heating rate of 10°C/min (first heating scan). The temperature was then lowered to 30°C at a heating rate of 10°C/min, and the samples submitted to a second heating scan under the same conditions as the first. Crystallinity degree was calculated using Eq. (1).

$$\% \text{ Cristalinidade} = \frac{\Delta H_f}{\Delta H_{100\%}} \text{ x } 100 \quad (1)$$

where ΔHm is the endothermic enthalpy, ΔH100% the theoretical melting enthalpy of 100% crystalline PP (209 J/g) [56], and wt% the amount of PP in the blend or composite.

The cryogenic-fractured surface morphology of the materials was examined under a scanning electron microscope (SEM, FEI, Quanta 400, accelerating voltage at 25 kV, 800X). The fractured samples were coated with gold.

3.4 Statistical analysis

The statistical analysis of the results was performed using STATISTICA 6 software. Analysis of variance (ANOVA) was applied to test for significant differences between the means. Residual normality and homogeneity of variances (Cochran C and Bartlett methods) were determined before univariate tests of significance and Fisher's least significant difference (LSD) test, using a significance level of $\alpha = 0.05$.

4. Results and discussion

4.1 Morphological and thermal characterization

Figure 2 shows the cryogenic fractured specimen for comparison between PP, PPr1/GP, and PPr3/GP. Poor interaction can be observed between the polymer matrix and glass particles due to the presence of small voids, gaps, and unattached particles.

The addition of PPMA in GP/PPr composites resulted in the smoothing of fractured surfaces (**Figure 3**). Furthermore, unattached particles, micro-voids and the gap between the matrix and the filler were slightly reduced. As such, filler-matrix interaction improved due to the addition of the coupling agent. **Table 3** shows the thermal properties of composites containing GP.

The melting temperatures (Tm) of all samples were similar to those of neat PP, except for the presence of a small endothermic peak in some composites at 127°C. Based on literature data [57], the small peak at 127°C can be attributed to the polyethylene and contamination in PPr, which is very common due to the difficulty in separating PP from PE during the recycling process.

In general, the composites showed similar crystallinity degrees (χc) to those of neat PP (**Table 3**). The results are noteworthy because they suggest that the presence of post-consumer materials (PPr and GP) did not disturb the crystal formation of the final composite, which leads to the assumption that the final properties of the composite are maintained, even with the addition of post-consumer materials.

Figure 2.
SEM micrographs of neat PP, PPr1/GP (90/10%w/w) and PPr3/GP (70/30%w/w) composites.

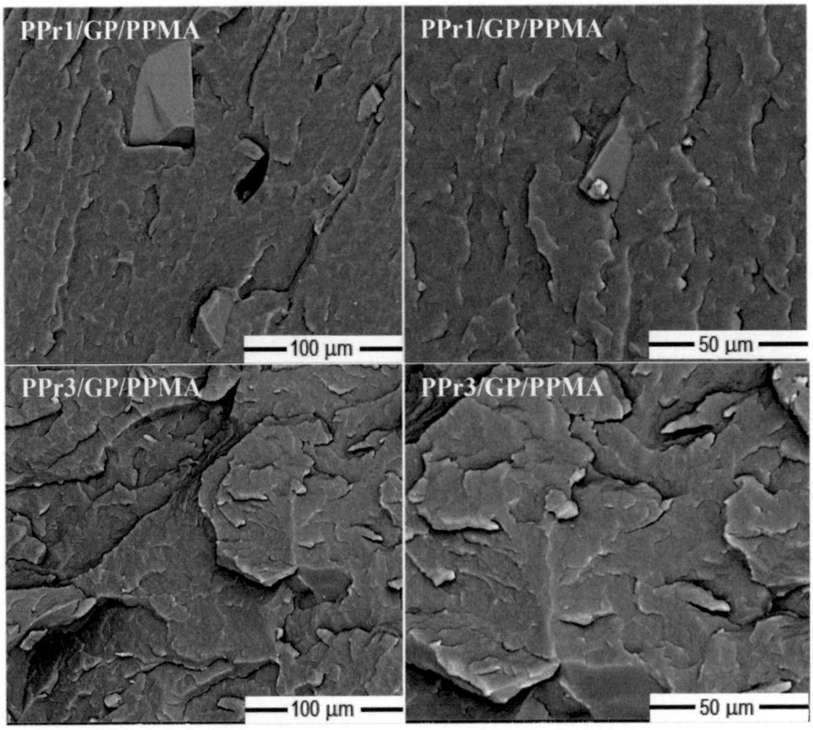

Figure 3.
SEM micrographs in two sizes (100 and 50 µm) of PP/PPr1/GP/PPMA (75/10/5/10%w/w/w/w) and PPr3/GP (55/30/5/10%w/w/w/w) composites.

Property	PP	PPr1/GP	PPr3/GP	PPr1/GP/PPMA	PPr3/GP/PPMA
Tm (°C)	166	165	127/165	126/165	126/164
ΔHm (J/g)	81.6	88.0	72.6	68.8	72.6
χc (%)	39	44	37	35	37

Table 3.
Thermal properties of PP and PP/PPr/GP-base composites.

4.2 Mechanical properties

Material tensile properties are important for both engineering and packaging applications, since they represent the ability of materials to withstand the load transferred in the longitudinal direction. Impact strength is essential to engineering applications, due to the need to bear high loads for very short periods of time. Thus, tensile properties are vital in evaluating bags or mooring ropes, while impact strength is a critical feature in recycled plastic fencing, furniture and automobile parts. Tensile properties, mean values and confidence intervals (Fisher's LSD test) are shown in **Figures 4–7**.

With respect to elastic modulus (**Figure 4**), a comparison between PPr1, PPr3, and PP showed that elastic modulus, which is related to composite rigidity, tends to decrease with the addition of PPr, but this effect is only significant for the composite with 30%wt of PPr. However, when recycled filler (GP) was added to PP/PPr blends, the modulus rose until the mean composite modulus values were equal to those of virgin PP. This finding suggests that the previous decline in stiffness observed in PP/PPr blends can be solved by adding glass powder.

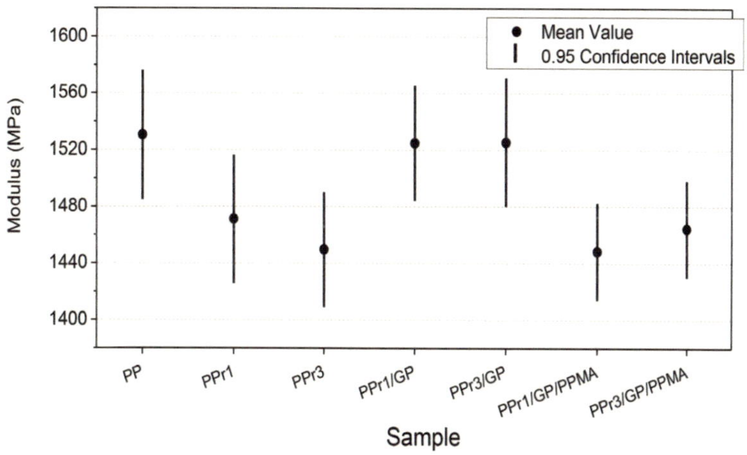

Figure 4.
Elastic modulus of PP and PP/PPr/GP-base composites.

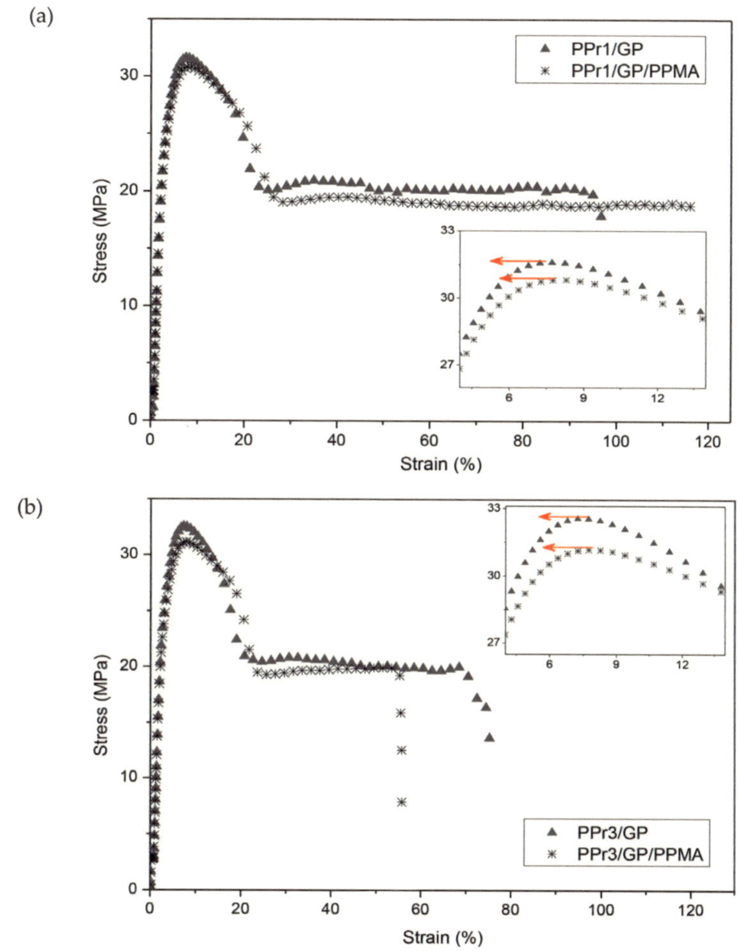

Figure 5.
Stress x strain curves of composites samples with: (a) 0% and (b) 10% of PPMA.

By contrast, the improvement in elastic modulus was reversed when PPMA was added to PP/PPr/GP composites. This is significant because good cohesion between the matrix and GP, as shown in **Figure 3**, was expected to improve the

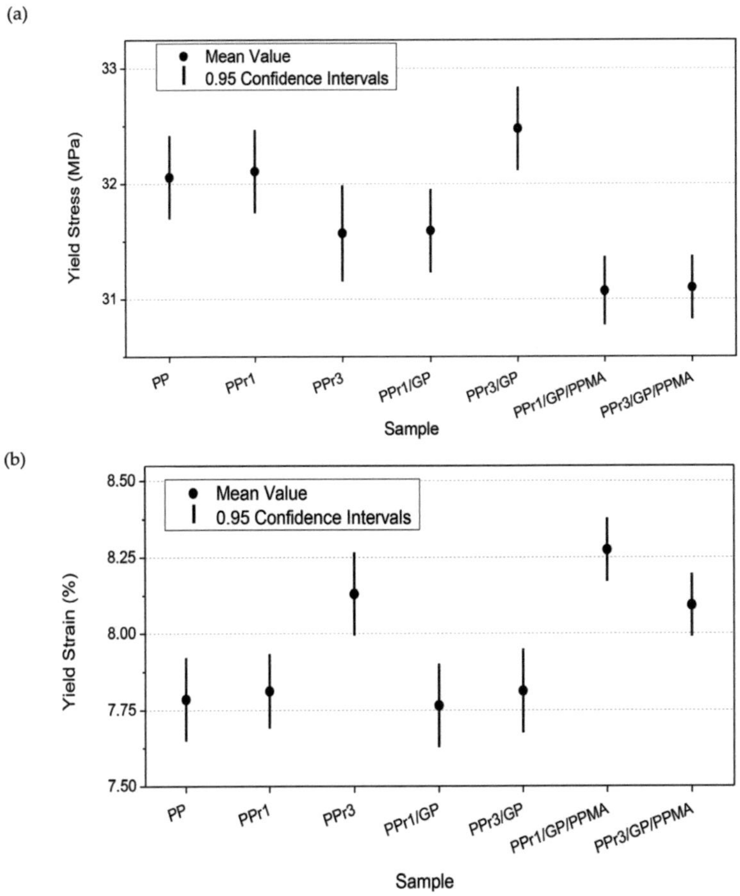

Figure 6.
Yield properties of PP and PP/PPr/GP-base composites: (a) yield stress and (b) yield strain.

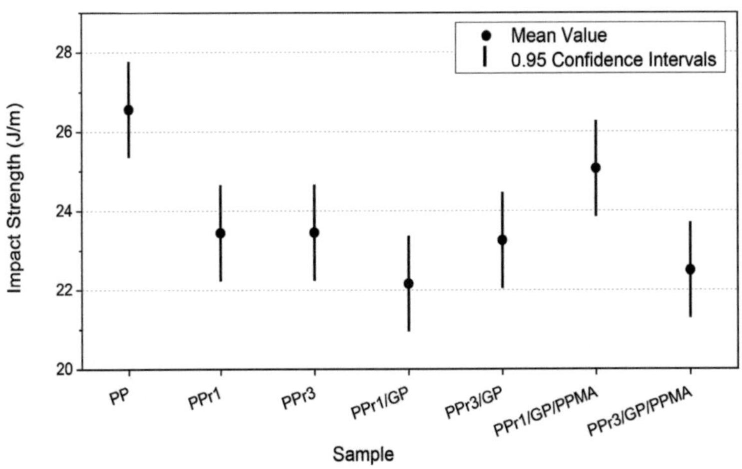

Figure 7.
Impact strength of PP and PP/PPr/GP-base composites.

tmodulus of the composite. One reason that may explain the previous undesirable result is the high amount of PPMA used as plasticizer. In other words, the amount of PPMA exceeded what was needed to coat the particle surfaces, diffusing in the polymer matrix and influencing plasticizer properties or those of a third polymer component.

Yield stress is the maximum stress at which the material begins to exhibit permanent deformation. As the elastic limit shifts, the material does not return to its original dimensions after the applied stress is removed. This property is particularly important for automotive applications. The yield stress of composites is illustrated on median stress x strain curves (**Figure 5**), showing the PPMA effect.

Yield properties (**Figure 6**) are generally in accordance with the modulus trend, that is, the higher the stiffness, the greater the stress and lower the strain at the yield point of the material. This behavior became evident when virgin PP was compared to PPr/GP composites. For products whose performance is highly dependent on tensile properties, the PPr3/GP sample remains the best option, considering elastic modulus, yield and environmental aspects, since the properties are very similar to those of neat PP even with the addition of 30% PPr and 5%wt of GP.

With respect to the automotive applications of plastic materials, acceptable impact strength is one of the requirements and, in the case of composites, this property is highly sensitive to particle/matrix debonding during mechanical energy dissipation. **Figure 7** shows the mean values and confidence intervals of impact strength.

The PPr1/GP/PPMA sample showed somewhat better results compared to virgin PP, with a P-value of 0.0838, indicating no significant difference between results. The use of PPMA as a coupling agent for GP could be optimized to improve the impact strength of composites.

In the present study, the PPMA grade used exhibited a maleic anhydride level of 1.0% by weight, but other grades with higher levels and greater affinity to GP could be tested in future research.

Compared to virgin resin, blends of PP with PPr demonstrated poor impact properties (**Figure 7**). In addition to higher impurity levels, PPr is expected to show lower molecular weight than PP, and both factors can contribute to failure in recycled material blends. Fukuhara et al. [58] evaluated isotactic polypropylene with different molecular weight and observed less Izod impact strength in samples with lower molecular weights. Furthermore, any structural particularity in PPr able to influence PP crystalline morphology can modify mechanical properties. Xu et al. [59] studied the relationship between spherulite size and crystallinity in the impact strength of PP samples with several different nucleating agents. The authors reported that impact strength was primarily controlled by spherulite size for samples with low crystallinity. For high crystalline samples, crystallinity itself is the decisive factor in strength. The authors also observed that impact strength is greater in PP samples with small spherulites and lower crystallinity. Nevertheless, no clear relationship between the degree of crystallinity and impact strength of samples was observed in the present study (**Table 3**, **Figure 7**). As such, other factors, such as impurity content, may exert the greatest influence on impact results. Given that products such as furniture and automobile parts require high impact strength, suitable coupling agents should be added to recycled composites in order to enhance their properties.

5. Conclusions

The greatest challenge to plastics in the automotive industry is in recycling. Some automotive manufacturers, such as Ford and Toyota, are recycling their vehicles plastics and reusing in the new vehicles, for example, old or damaged bumpers are recycled and reused in bumper reinforcement cores [60].

According to the results presented, it can be concluded that the properties did not vary significantly as a function of composition. Since the objective was to produce lower cost composites (incorporating recycled PP and glass powder) and more sustainable materials without significant loss of properties, this result is within the parameters established. In other words, it was possible to recover post-consumer materials, replacing the virgin resin without significant loss of mechanical integrity in the final product. The addition of maleated polypropylene (PPMA) was shown to significantly improve the toughness of the material.

In conclusion, based on the properties analyzed and the sustainable appeal of the new products, the powder-based composites displayed potential for use in various applications in the automotive industry, replacing conventional materials.

Acknowledgements

The authors would like to thank Massfix company for donating the glass powder (GP), Poli Injet company for supplied the post-industrial polypropylene (PPr) and Braskem company for donating the virgin polypropylene, and PIBIC/UFRJ for the scholarship grants.

Conflict of interest

The authors declare that there is no conflict of interest regarding the publication of this paper.

Author details

Denis R. Dias[1], Maria José O. C. Guimarães[1], Christine R. Nascimento[2], Celio A. Costa [2], Giovanio L. de Oliveira[2], Mônica C. de Andrade[3], Ana Maria F. de Sousa[4], Ana Lúcia N. da Silva[5,6] and Elen B. A. Vasques Pacheco[5,6]*

1 Departamento de Processos Orgânicos, Escola de Química, Universidade Federal do Rio de Janeiro, Rio de Janeiro, Brazil

2 Programa de Engenharia Metalúrgica e de Materiais, Universidade Federal do Rio de Janeiro, Rio de Janeiro, Brazil

3 Instituto Politécnico do Rio de Janeiro, Universidade do Estado do Rio de Janeiro, Nova Friburgo, Brazil

4 Instituto de Química, Universidade do Estado do Rio de Janeiro, Rio de Janeiro, Brazil

5 Instituto de Macromoléculas Professora Eloisa Mano, Universidade Federal do Rio de Janeiro, Rio de Janeiro, Brazil

6 Programa de Engenharia Ambiental, Universidade Federal do Rio de Janeiro, Rio de Janeiro, Brazil

*Address all correspondence to: elen@ima.ufrj.br

References

[1] Kirchherr J, Reike D, Hekkert M. Conceptualizing the circular economy: An analysis of 114 definitions. Resources, Conservation and Recycling. 2017;**127**:221-232. DOI: 10.1016/j.resconrec.2017.09.005

[2] Michelini G, Moraes RN, Cunha RN, Costa JMH, Ometto AR. From linear to circular economy: PSS conducting the transition. Procedia CIRP. 2017;**64**:2-6. DOI: 10.1016/j.procir.2017.03.012

[3] Korhonen J, Honkasal JS. Circular economy: The concept and its limitations. Ecological Economics. 2018;**143**:37-46. DOI: 10.1016/j.ecolecon.2017.06.041

[4] Ritzén S, Sandström GÖ. Barriers to the circular economy—Integration of perspectives and domains. Procedia CIRP. 2017;**64**:7-12. DOI: 10.1016/j.procir.2017.03.005

[5] Ghisellini P, Cialani C, Ulgiati S. A review on circular economy: The expected transition to a balanced interplay of environmental and economic systems. Journal of Cleaner Production. 2016;**114**:11-32. DOI: 10.1016/j.jclepro.2015.09.007

[6] Pomponi F, Moncaster A. Circular economy for the built environment: A research framework. Journal of Cleaner Production. 2017;**143**:710-718. DOI: 10.1016/j.jclepro.2016.12.055

[7] North EJ, Halden RU. Plastics and environmental health: The road ahead. Reviews on Environmental Health. 2013;**28**(1):1-8. DOI: 10.1515/reveh-2012-0030

[8] Andrady AL, Neal MA. Applications and societal benefits of plastics. Philosophical Transactions of the Royal Society of London. Series B, Biological Sciences. 2009;**364**(1526):1977-1984. DOI: 10.1098/rstb.2008.0304

[9] Geyer R, Jambeck JR, Law KL. Production, use, and fate of all plastics ever made. Science Advances. 2017;**3**:1-5. DOI: 10.1126/sciadv.1700782

[10] Bernardo CA, Simões C, Pinto LMC. Environmental and economic life cycle analysis of plastic waste anagement options. A review. In: Proceedings of the Regional Conference Graz 2015—Polymer Processing Society PPS: Conference Papers; 2016. DOI: 10.1063/1.4965581

[11] Duval D, MacLean HL. The role of product information in automotive plastics recycling: A financial and life cycle assessment. Journal of Cleaner Production. 2007;**15**:1158-1168. DOI: 10.1016/j.jclepro.2006.05.030

[12] Arena U, Mastellone ML, Perugini F. Life cycle assessment of a plastic packaging recycling system. The International Journal of Life Cycle Assessment. 2003;**8**(2):92-98. DOI: 10.1065/Ica2003.02.106

[13] Rajendran S, Hodzic A, Soutis C, Al-Maadeed MA. Review of life cycle assessment on polyolefins and related materials. Plastics, Rubber and Composites. 2012;**41**(4/5):159-168. DOI: 10.1179/1743289811Y.0000000051

[14] Horodytska O, Valdés FJ, Fullana A. Plastic flexible films waste management—A state of art review. Waste Management. 2018;**77**:413-415. DOI: 10.1016/j.wasman.2018.04.023

[15] Conceição RDP, Pereira C, Pessoa G, Pacheco EBAV. The post-consumer PET recycling chain and definition of their stages: A case study in Rio de Janeiro. Revista Brasileira de Ciências Ambientais. 2016;**39**:80-96. DOI: 10.5327/Z2176-9478201613514

[16] Sinha V, Patel MR, Patel JV. PET waste management by chemical

recycling: A review. Journal of Polymers and the Environment. 2010;**18**(1):8-25. DOI: 10.1007/s10924-008-0106-7

[17] Huysman S, Debaveye S, Schaubroeck T, DeMeester S, Ardente F, Mathieux F, et al. The recyclability benefit rate of closed-loop and open-loop systems: A case study on plastic recycling in Flanders. Resources, Conservation and Recycling. 2015;**101**:53-60. DOI: https://doi.org/10.1016/j.resconrec.2015.05.014

[18] Chilton T, Burnley S, Nesaratnam S. A life cycle assessment of the closed-loop recycling and thermal recovery of post-consumer PET. Resources, Conservation and Recycling. 2010;**54**(12):1241-1249. DOI: https://doi.org/10.1016/j.resconrec.2010.04.002

[19] Williams TGJL, Heidrich O, Sallis P. A case study of the open-loop recycling of mixed plastic waste for use in a sportsfield drainage system. Resources, Conservation and Recycling. 2010;**55**(2):118-128. DOI: 10.1016/j.resconrec.2010.08.002

[20] Shen L, Worrel E, Patel MK. Open-loop recycling: A LCA case study of PET bottle-to-fibre recycling. Resources, Conservation and Recycling. 2010;**55**(1):34-52. DOI: 10.1016/j.resconrec.2010.06.014

[21] Fernandes BL, Domingues AJ. Mechanical characterization of recycled polypropylene for automotive industry. Revista Polímeros: Ciência e Tecnologia. 2007;**7**:85-87. DOI: 10.1590/S0104-14282007000200005

[22] Hemais CA. Polymers and the automobile industry. Revista Polímeros: Ciência e Tecnologia. 2003;**13**:107-114. DOI: 10.1590/S0104-14282003000200008

[23] Maddah HA. Polypropylene as a promising plastic: A review. American Journal of Polymer Science. 2016;**6**(1):1-11. DOI: 10.5923/j.ajps.20160601.01

[24] Moritomi S, Watanabe T, KanzakiI S. Polypropylene Compounds for Automotive Applications [Internet]. 2010. Translated from R&D Report, "SUMITOMO KAGAKU". Vol. 2010-I. Available from: http://www.sumitomo-chem.co.jp/english/rd/report/theses/docs/20100100_a2g.pdf [Accessed: 09-05-2018]

[25] Santos LS, Silva AHMFT, Pacheco EBAV, Silva ALN. Avaliação do efeito da adição de PP reciclado nas propriedades mecânicas e de escoamento de misturas PP/EPDM. Revista Polímeros: Ciência e Tecnologia. 2013;**23**(3):389-394. DOI: 10.4322/polímeros.2013.083

[26] Da Silva Spinacé MA, De Paoli MA. The technology of polymer recycling. Química Nova. 2005;**28**:65-72. DOI: 10.1590/S0100-40422005000100014

[27] Valera TS, Sakai ACV, Wiebeck H, Toffoli SM. Propriedades do compósito poliamida-6/vidro em pó. Anais Congresso Brasileiro de Engenharia e Ciência dos Materiais, São Pedro/SP; n°14. 2000. pp. 49401-49411. Available from: https://www.ipen.br/biblioteca/cd/cbecimat/2000/Docs/TC403-010.pdf [Accessed: 25-08-2018]

[28] Medina HV, Gomes DA. The automobile industry designing for recycling. In: CETEM, 2002. Available from: http://web-resol.org/textos/ferro_reciclagem_automoveis_brasil.pdf [Accessed: 25-08-2018]

[29] Swain B, Ryang Park J, Yoon Shin D, Park KS, Hwan Hong M, Gi Lee C. Recycling of waste automotive laminated glass and valorization of polyvinil butyral through mechanochemical separation. Environmental Research. 2015;**142**:615-623. DOI: 10.1016/j.envres.2015.08.017

[30] Sheril RV, Mariatti M, Samayamutthirian P. Single and hybrid mineral fillers (talc/silica and talc/calcium carbonate)-filled polypropylene composites: Effects of filler loading and ratios. Journal of Vinyl and Additive Technology. 2014;**20**:160-167. DOI: 10.1002/vnl.21347

[31] Yu M, Huang R, He C, Wu Q, Zhao XD, Hom N. Hybrid composites from wheatstraw, inorganic filler, and recycled polypropylene: Morphology and mechanical and thermal expansion performance. International Journal of Polymer Science. 2016;**2016**:1-12. ID2520670. DOI: 10.1155/2016/2520670

[32] Srivabut C, Ratanawilai T, Hiziroglu S. Effect of nanoclay, talcum, and calcium carbonate as filler on properties of composites manufactured from recycled polypropylene and rubberwood fiber. Construction and Bulding Materials. 2018;**162**:450-458. DOI: 10.1016/j.conbuildmat.2017.12.048

[33] Rabello M. Aditivação de Polímeros, Editora Artliber. São Paulo: ABPol; 2011

[34] Fernandes BL, Domingues AJ. Mechanical characterization of recycled polypropylene for automotive industry. Polímeros. 2007;**17**:85-87. DOI: 10.1590/S0104-14282007000200005

[35] Ota WN. Análise de compósitos de polipropileno e fibras de vidro utilizados pela indústria automotiva nacional [Dissertation]. Curitiba: Universidade Federal do Paraná; 2004. 90 f. Available from: http://www.pipe.ufpr.br/portal/defesas/dissertacao/058.pdf [Accessed: 25-08-2018]

[36] Kozderka M, Rose B, Bahlouli N, Kocí V, Caillaud E. Recycled high impact polypropylene in the automotive indudtry-mechanical and environmental properties. International Journal on Interactive Design and Manufacturing. 2017;**11**:737-750. DOI: 10.1007/s12008-016-0365-9

[37] Yu Y, Yang Y, Murakami M, Nomura M, Hamada H. Physical and mechanical properties of injection-molded wood powder thermoplastic composites. Advanced Composite Materials. 2013;**22**:425-435. DOI: 10.1080/09243046.2013.843825

[38] Nacas AM, Silva RL, De Paoli MA, Spinacé MAS. Polypropylene composite reinforced with fibrillated curaua fiber and using maleic anhydride as coupling agent. Journal of Applied Polymer Science. 2017;**134**:1-8. DOI: 10.1002/APP.44913

[39] Xanthos M. Recycling of the #5 polymer. Science (New York, N.Y.). 2012;**337**:700-702. DOI: 10.1126/science.1221806

[40] Tan T, Santos SF, Savastano H Jr, Soboyejo WO. Fracture and resistance curve behavior in hybrid natural fiber and polypropylene fiber reinforced composites. Journal of Materials Science. 2012;**47**:2864-2874. DOI: 10.1007/s10853-011-6116-1

[41] Wu JH, Chen CW, Wu YT, Wu GT, Kuo MC, Tsai Y. Mechanical properties, morphology, and crystallization behavior of polypropylene/elastomer/talc composites. Polymer Composites. 2015:69-77. DOI: 10.1002/pc.22914

[42] Hanim H, Fuad MYA, Zarina R, Ishak ZAM, Hassan A. Properties and structure of polypropylene/polyethylene-octene elastomer/nano $CaCO_3$ composites. Journal of Thermoplastic Composite Materials. 2008;**21**:123-140. DOI: 10.1177/0892705707083634

[43] Goldman AY, Copsey CJ. Polypropylene toughened with calcium carbonate mineral filler. Materials Research Innovations. 2000;**3**:302-307. DOI: 10.1007/s100190000053

[44] Li XH, He Y, Li X, An F, Yang D, Yu ZZ. Simultaneous enhancements in toughness and electrical conductivity of polypropylene/carbon nanotube nanocomposites by incorporation of electrically inert carbonate nanoparticles. Industrial & Engineering Chemistry Research. 2017;**56**:2783-2788. DOI: 10.1021/acs-iecr.7b00446

[45] Iyer KA, Torkelson JM. Importance of superior dispersion versus filler surface modification in producing robust polymer nanocomposites: The example of polypropylene/nanosílica hybrids. Polymer. 2015;**68**:147-157. DOI: 10.1016/j.polymer.2015.05.015

[46] Vache SD, Plummer CJG, Houphouët-Boigny C, Manson JAE. Morphology and mechanical properties of isotactic polypropylene glass mat thermoplastic composites modified with organophilic montmorillonite. Journal of Materials Science. 2011;**46**:2112-2122. DOI: 10.1007/s10853-010-5047-6

[47] Zhou X, Gance D, Qunfang L. Improvement in impact property of continuous glass mat reinforced polypropylene composites. Journal of Applied Polymer Science. 2002;**83**:2680-2688. DOI: 10.1002/app.10245

[48] Matheus FL, Rawlings RD. Composite Materials: Engineering and Science. UK: Chapman and Hall; 1994

[49] Wambua P, Ivens J, Verpoest I. Natural Fibres: Can they Replace Glass in Fibre Reinforced Plastics? Belgium: Elsevier Science; 2003

[50] Kim T, Oshima K, Kawada H. Impact tensile properties and strength development mechanism of glass for reinforcement fiber. Journal of Physics: Conference Series. 2013;**451**:1-6. DOI: 10.1088/1742-6596/451/1/012006

[51] Thomason JL, Schoolemberg GE. An investigation of glass fibre/polypropylene interface strength and its effect on composite properties. Composites. 1994;**25**:197-203

[52] Chen Y, Wen X, Nie M, Wang Q. Preparation of polypropylene/glass fiber composite with high performance through interfacial crystallization. Journal of Vinyl & Additive Technology. 2017:284-289. DOI: 10.1002/vnl.21514

[53] Delogu M, Zanchi L, Maltese S, Bonoli A, Pierini M. Environmental and economic life cycle assessment of a lightweight solution for an automotive component: A comparison between talc-filled and hollow glass microspheres-reinforced polymer composites. Journal of Cleaner Production. 2016;**139**:548-560. DOI: 10.1016/j.jclepro.2016.08.079

[54] Antunes A. Setores da Indústria Química Orgânica. 2007. Available from: https://books.google.com.br/books?isbn=8576500779 [Accessed: 17-05-2018]

[55] Reis GF. Análises mecânicas da mistura polipropileno com material reciclado e com negro de fumo para o uso na indústria automobilística [graduation work]. Rio de Janeiro: Universidade Federal Fluminense; 2012

[56] Guo C, Ma L, Wang Q. Influence of compatibilizer and wood flour on the non-isothermal crystallization behavior of polypropylene composites. Polymers and Polymer Composites. 2010;**18**:37-44

[57] Santos LS. Avaliação da influência da incorporação de polipropileno pós-consumo em misturas de PP/EPDM para indústrias automobilisticas [thesis]. Instituto de Macromoléculas Professora Eloisa Mano: Universidade Federal do Rio de Janeiro; 2011

[58] Fukuhara N. Influence of molecular weight on J-integral testing of polypropylene. Polymer Testing. 1999;**18**:135-149. DOI: 10.1016/S0142-9418(98)00021-X

[59] Xu T, Yu J, Jin Z. Effects of crystalline morphology on the impact behavior of polypropylene. Materials and Design. 2001;**22**:27-31. DOI: 10.1016/S0261-3069(00)00033-9

[60] Srivastava V, Srivastava R. Advances in automotive polymer applications and recycling. International Journal of Innovative Research in Science, Engineering and Technology. 2013;**2**:744-746. Available from: https://www.ijirset.com/upload/march/36_Advances%20in.pdf [Accessed: 25-08-2018]

Chapter 5

Biological Degradation of Polymers in the Environment

John A. Glaser

Abstract

Polymers present to modern society remarkable performance characteristics desired by a wide range of consumers but the fate of polymers in the environment has become a massive management problem. Polymer applications offer molecular structures attractive to product engineers desirous of prolonged lifetime properties. These characteristics also figure prominently in the environmental lifetimes of polymers or plastics. Recently, reports of microbial degradation of polymeric materials offer new emerging technological opportunities to modify the enormous pollution threat incurred through use of polymers/plastics. A significant literature exists from which developmental directions for possible biological technologies can be discerned. Each report of microbial mediated degradation of polymers must be characterized in detail to provide the database from which a new technology developed. Part of the development must address the kinetics of the degradation process and find new approaches to enhance the rate of degradation. The understanding of the interaction of biotic and abiotic degradation is implicit to the technology development effort.

Keywords: polymers, plastics, degradation, microbial degradation, biofilms, extent of degradation

1. Introduction

In 1869, the first synthetic polymer was invented in response to a commercial $10,000 prize to provide a suitable replacement to ivory. A continuous string of discoveries and inventions contributed new polymers to meet the various requirements of society. Polymers are constructed of long chains of atoms, organized in repeating components or units often exceeding those found in nature. Plastic can refer to matter that is pliable and easily shaped. Recent usage finds it to be a name for materials called polymers. High molecular weight organic polymers derived from various hydrocarbon and petroleum materials are now referred to as plastics [1].

Synthetic polymers are constructed of long chains of smaller molecules connected by strong chemical bonds and arranged in repeating units which provide desirable properties. The chain length of the polymers and patterns of polymeric assembly provide properties such as strength, flexibility, and a lightweight feature that identify them as plastics. The properties have demonstrated the general utility of polymers and their manipulation for construction of a multitude of widely useful items leading to a world saturation and recognition of their unattractive properties too. A major trend of ever increasing consumption of plastics has been seen in the

 IntechOpen

Polymer	Abbreviation	Density (23/4°C)	Crystallinity (%)	Lifespan (year)
Polyethylene	PE	0.91–0.925	50	10–600
Polypropylene	PP	0.94–0.97	50	10–600
Polystyrene	PS	0.902–0.909	0	50–80
Polyethylene glycol terephthalate	PET	1.03–1.09	0–50	450
Polyvinyl chloride	PVC	1.35–1.45	0	50–100+

Table 1.
Selected features of major commercial thermoplastic polymers [7].

areas of industrial and domestic applications. Much of this polymer production is composed of plastic materials that are generally non-biodegradable. This widespread use of plastics raises a significant threat to the environment due to the lack of proper waste management and a until recently cavalier community behavior to maintain proper control of this waste stream. Response to these conditions has elicited an effort to devise innovative strategies for plastic waste management, invention of biodegradable polymers, and education to promote proper disposal. Technologies available for current polymer degradation strategies are chemical, thermal, photo, and biological techniques [2–6]. The physical properties displayed in **Table 1** show little differences in density but remarkable differences in crystallinity and lifespan. Crystallinity has been shown to play a very directing role in certain biodegradation processes on select polymers.

Polymers are generally carbon-based commercialized polymeric materials that have been found to have desirable physical and chemical properties in a wide range of applications. A recent assessment attests to the broad range of commercial materials that entered to global economy since 1950 as plastics. The mass production of virgin polymers has been assessed to be 8300 million metric tons for the period of 1950 through 2015 [8]. Globally consumed at a pace of some 311 million tons per year with 90% having a petroleum origin, plastic materials have become a major worldwide solid waste problem. Plastic composition of solid waste has increased for less than 1% in 1960 to greater than 10% in 2005 which was attributed largely to packaging. Packaging plastics are recycled in remarkably low quantities. Should current production and waste management trends continue, landfill plastic waste and that in the natural environment could exceed 12,000 Mt of plastic waste by 2050 [9].

2. Polymer structures and features

A polymer is easily recognized as a valuable chemical made of many repeating units [10]. The basic repeating unit of a polymer is referred to as the "-mer" with "poly-mer" denoting a chemical composed of many repeating units. Polymers can be chemically synthesized in a variety of ways depending on the chemical characteristics of the monomers thus forming a desired product. Nature affords many examples of polymers which can be used directly or transformed to form materials required by society serving specific needs. The polymers of concern are generally composed of carbon and hydrogen with extension to oxygen, nitrogen and chlorine functionalities (see **Figure 1** for examples). Chemical resistance, thermal and electrical insulation, strong and light-weight, and myriad applications where no alternative exists are polymer characteristics that continue to make polymers attractive. Significant polymer application can be found in the automotive, building and construction, and packaging industries [12].

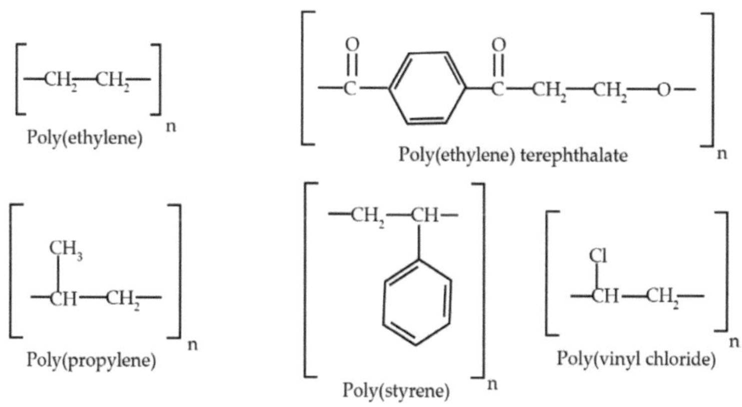

Figure 1.
Structures of major commercial thermoplastic polymers [11].

The environmental behavior of polymers can be only discerned through an understanding of the interaction between polymers and environment under ambient conditions. This interaction can be observed from surface properties changes that lead to new chemical functionality formation in the polymer matrix. New functional groups contribute to continued deterioration of the polymeric structure in conditions such as weathering. Discoloration and mechanical stiffness of the polymeric mass are often hallmarks of the degradative cycle in which heat, mechanical energy, radiation, and ozone are contributing factors [13].

Polyolefins (PO) are the front-runners of the global industrial polymer market where a broad range of commercial products contribute to our daily lives in the form o packaging, bottles, automobile parts and piping. The PO class family is comprised of saturated hydrocarbon polymers such as high-density polyethylene (HDPE), low-density polyethylene (LDPE) and linear low-density polyethylene (LLDPE), propylene and higher terminal olefins or monomer combinations as copolymers. The sources of these polymers are low-cost petrochemicals and natural gas with monomers production dependent on cracking or refining of petroleum. This class of polymers has a unique advantage derived from their basic composition of carbon and hydrogen in contrast to other available polymers such as polyurethanes, poly(vinyl chloride) and polyamides [14].

The copolymers of ethylene and propylene are produced in quantities that exceed 40% of plastics produced per annum with no production leveling in sight. This continuous increase suggests that as material use broadens yearly, the amount of waste will also increase and present waste disposal problems. Polyolefin biological and chemical inertness continues to be recognized as an advantage. However, this remarkable stability found at many environmental conditions and the degradation resistance leads to environmental accumulation and an obvious increase to visible pollution and ancillary contributing problems. Desired environmental properties impact the polyolefin market on the production side as well as product recyclability [15].

3. Biological degradation

Biodegradation utilizes the functions of microbial species to convert organic substrates (polymers) to small molecular weight fragments that can be further degraded to carbon dioxide and water [16–21]. The physical and chemical properties of a polymer are important to biodegradation. Biodegradation efficiency achieved by the microorganisms is directly related to the key properties such as molecular weight and crystallinity of the polymers. Enzymes engaged in polymer degradation initially are outside the cell and are referred to as exo-enzymes having a wide reactivity ranging from oxidative to hydrolytic functionality. Their action on the polymer can be generally described as depolymerization. The exo-enzymes generally degrade complex polymer structure to smaller, simple units that can take in the microbial cell to complete the process of degradation.

3.1 Requirements to assay polymer biodegradation

Polymer degradation proceeds to form new products during the degradation path leading to mineralization which results in the formation of process end-products such as, e.g., CO_2, H_2O or CH_4 [22]. Oxygen is the required terminal electron acceptor for the aerobic degradation process. Aerobic conditions lead to the formation of CO_2 and H_2O in addition to the cellular biomass of microorganisms during the degradation of the plastic forms. Where sulfidogenic conditions are found, polymer biodegradation leads to the formation of CO_2 and H_2O. Polymer degradation accomplished under anaerobic conditions produces organic acids, H_2O, CO_2, and CH_4. Contrasting aerobic degradation with anaerobic conditions, the aerobic process is found to be more efficient. When considering energy production the anaerobic process produces less energy due to the absence of O_2, serving the electron acceptor which is more efficient in comparison to CO_2 and SO_4^{-2} [23].

As solid materials, plastics encounter the effects of biodegradation at the exposed surface. In the unweathered polymeric structure, the surface is affected by biodegradation whereas the inner part is generally unavailable to the effects of biodegradation. Weathering may mechanically affect the structural integrity of the plastic to permit intrusion of bacteria or fungal hyphae to initiate biodegradation at inner loci of the plastic. The rate of biodegradation is functionally dependent on the surface area of the plastic. As the microbial-colonized surface area increases, a faster biodegradation rate will be observed assuming all other environmental conditions to be equal [24].

Microorganisms can break organic chemicals into simpler chemical forms through biochemical transformation. Polymer biodegradation is a process in which any change in the polymer structure occurs as a result of polymer properties alteration resulting from the transformative action of microbial enzymes, molecular weight reduction, and changes to mechanical strength and surface properties attributable to microbial action. The biodegradation reaction for a carbon-based polymer under aerobic conditions can be formulated as follows:

$$C_{Polymer} + O_2 + \text{Biomass} \longrightarrow CO_2 + H_2O + C_{biomass} \qquad (1)$$

Assimilation of the carbon comprising the polymer ($C_{polymer}$) by microorganisms results in conversion to CO_2 and H_2O with production of more microbial biomass ($C_{biomass}$). In turn, $C_{biomass}$ is mineralized across time by the microbial community or held in reserve as storage polymers [25].

The following set of equations is a more complete description of the aerobic plastic biodegradation process:

$$\begin{array}{l} C_{Polymer} + O_2 \longrightarrow \text{Oligomers} + CO_2 + H_2O + C_{biomass} \\ \qquad\qquad\qquad\qquad \searrow \\ \qquad\qquad\qquad\qquad\quad C_{biomass} + H_2O + CO_2 \end{array} \qquad (2)$$

where $C_{polymer}$ and newly formed oligomers are converted into $C_{biomass}$ but $C_{biomass}$ converts to CO_2 under a different kinetics scheme. The conversion to CO_2 is referred to as microbial mineralization. Each oligomeric fragment is expected to proceed through of sequential steps in which the chemical and physical properties are altered leading to the desired benign result. A technology for monitoring aerobic biodegradation has been developed and optimized for small organic pollutants using oxygen respirometry where the pollutant degrades at a sufficiently rapid rate for respirometry to provide expected rates of biodegradation. When polymers are considered, a variety of analytical approaches relating to physical and chemical changes are employed such as differential scanning calorimetry, scanning electron microscopy, thermal gravimetric analysis, Fourier transform infrared spectrometry, gas chromatograph-mass spectrometry, and atomic force microscopy [26].

Since most polymer disposal occurs in our oxygen atmosphere, it is important to recognize that aerobic biodegradation will be our focus but environmental anaerobic conditions do exist that may be useful to polymer degradation. The distinction between aerobic and anaerobic degradation is quite important since it has been observed that anaerobic conditions support slower biodegradation kinetics. Anaerobic biodegradation can occur in the environment in a variety of situations. Burial of polymeric materials initiates a complex series of chemical and biological reactions. Oxygen entrained in the buried materials is initially depleted by aerobic bacteria. The following oxygen depleted conditions provide conditions for the initiation of anaerobic biodegradation. The buried strata are generally covered by 3-m-thick layers which prevent oxygen replenishment. The alternate electron acceptors such as nitrate, sulfate, or methanogenic conditions enable the initiation of anaerobic biodegradation. Any introduction of oxygen will halt an established anaerobic degradation process.

3.2 Formulation of newer biodegradation schema

This formulation for the aerobic biodegradation of polymers can be improved due to the complexity of the processes involved in polymer biodegradation [27]. Biodegradation, defined as a decomposition of substances by the action of microorganisms, leading to mineralization and the formation of new biomass is not conveniently summarized. A new analysis is necessary to assist the formulation of comparative protocols to estimate biodegradability. In this context, polymer biodegradation is defined as a complex process composed of the stages of biodeterioration, biofragmentation, and assimilation [28].

The biological activity inferred in the term biodegradation is predominantly composed of, biological effects but within nature biotic and abiotic features act synergistically in the organic matter degradation process. Degradation modifying mechanical, physical and chemical properties of a material is generally referred to as deterioration. Abiotic and biotic effects combine to exert changes to these properties. This biological action occurs from the growth of microorganisms on the polymer surface or inside polymer material. Mechanical, chemical, and enzymatic means are exerted by microorganisms, thereby modifying the gross polymer material properties. Environmental conditions such as atmospheric pollutants, humidity, and weather strongly contribute to the overall process. The adsorbed pollutants can assist the material colonization by microbial species. A diverse collection of bacteria, protozoa, algae, and fungi are expected participants involved in biodeterioration. The development of different biota can increase biodeterioration by facilitating the production of simple molecules.

Fragmentation is a material breaking phenomenon required to meet the constraints for the subsequent event called assimilation. Polymeric material has a high molecular weight which is restricted by its size in its transit across the cell wall or cytoplasmic membrane. Reduction of polymeric molecule size is indispensable to this process. Changes to molecular size can occur through the involvement of abiotic and biotic processes which are expected to reduce molecular weight and size. The utility of enzymes derived from the microbial biomass could provide the required molecular weight reductions. Mixtures of oligomers and/or monomers are the expected products of the biological fragmentation.

Assimilation describes the integration of atoms from fragments of polymeric materials inside microbial cells. The microorganisms benefit from the input of energy, electrons and elements (i.e., carbon, nitrogen, oxygen, phosphorus, sulfur and so forth) required for the cell growth. Assimilated substrates are expected to be derived from biodeterioration and biofragmentation effects. Non-assimilated materials, impermeable to cellular membranes, are subject to biotransformation reactions yielding products that may be assimilated. Molecules transported across the cell membrane can be oxidized through catabolic pathways for energy storage and structural cell elements. Assimilation supports microbial growth and reproduction as nutrient substrates (e.g., polymeric materials) are consumed from the environment.

3.3 Factors affecting biodegradability

The polymer substrate properties are highly important to any colonization of the surface by either bacteria or fungi [29]. The topology of the surface may also be important to the colonization process. The polymer properties of molecular weight, shape, size and additives are each unique features which can limit biodegradability. The molecular weight of a polymer can be very limiting since the microbial colonization depends on surface features that enable the microorganisms to establish a locus from which to expand growth. Polymer crystallinity can play a strong role since it has been observed that microbial attachment to the polymer surface occurs and utilizes polymer material in amorphous sections of the polymer surface. Polymer additives are generally low molecular weight organic chemicals that can provide a starting point for microbial colonization due to their ease of biodegradation (**Figure 2**).

Weather is responsible for the deterioration of most exposed materials. Abiotic contributors to these conditions are moisture in its variety of forms, non-ionizing radiation, and atmospheric temperature. When combined with wind effects, pollution, and atmospheric gases, the overall process of deterioration can be quite formable. The ultraviolet (UV) component of the solar spectrum contributes ionizing radiation which plays a significant role in initiating weathering effects. Visible and near-infrared radiation can also contribute to the weathering process. Other factors

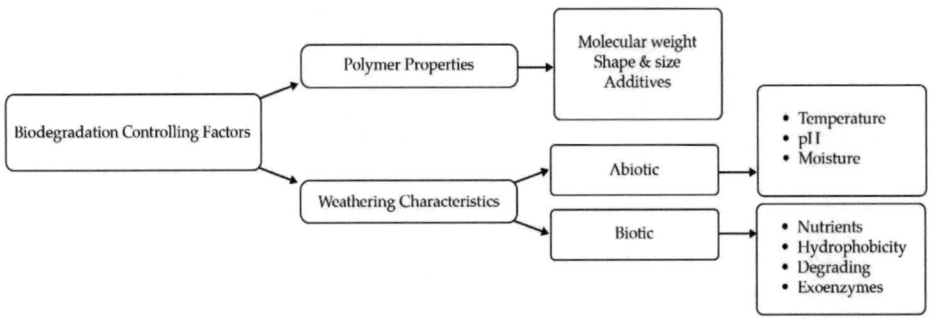

Figure 2.
Factors controlling polymer biodegradation [30].

couple with solar radiation synergistically to significantly influence the weathering processes. The quality and quantity of solar radiation, geographic location changes, time of day and year, and climatological conditions contribute to the overall effects. Effects of ozone and atmospheric pollutants are also important since each can interact with atmospheric radiation to result in mechanical stress such as stiffening and cracking. Moisture when combined with temperature effects can assist microbial colonization. The biotic contributors can strongly assist the colonization by providing the necessary nutrients for microbial growth. Hydrophilic surfaces may provide a more suitable place for colonization to ensue. Readily available exoenzymes from the colonized area can initiate the degradation process.

3.4 Biofilms

Communities of microorganisms attached to a surface are referred to as biofilms [31]. The microorganisms forming a biofilm undergo remarkable changes during the transition from planktonic (free-swimming) biota to components of a complex, surface-attached community (**Figure 3**). The process is quite simple with planktonic microorganism encountering a surface where some adsorb followed by surface release to final attachment by the secretion of exopolysaccharides which act as an adhesive for the growing biofilm [33]. New phenotypic characteristics are exhibited by the bacteria of a biofilm in response to environmental signals. Initial cell-polymer surface interactions, biofilm maturation, and the return to planktonic mode of growth have regulatory circuits and genetic elements controlling these diverse functions. Studies have been conducted to explore the genetic basis of biofilm development with the development of new insights. Compositionally, these films have been found to be a single microbial species or multiple microbial species with attachment to a range of biotic and abiotic surfaces [34, 35]. Mixed-species biofilms are generally encountered in most environments. Under the proper nutrient and carbon substrate supply, biofilms can grow to massive sizes. With growth, the biofilm can achieve large film structures that may be sensitive to physical forces such as agitation. Under such energy regimes, the biofilm can detach. An example of biofilm attachment and utility can be found in the waste water treatment sector where large polypropylene disks are rotated through industrial or agriculture waste water and then exposed to the atmosphere to treat pollutants through the intermediacy of cultured biofilms attached to the rotating polypropylene disk.

Biofilm formation and activity to polymer biodegradation are complex and dynamic [36]. The physical attachment offers a unique scenario for the attached microorganism and its participation in the biodegradation. After attachment as a biofilm component, individual microorganisms can excrete exoenzymes which can provide a range of functions. Due to the mixed-species composition found in most

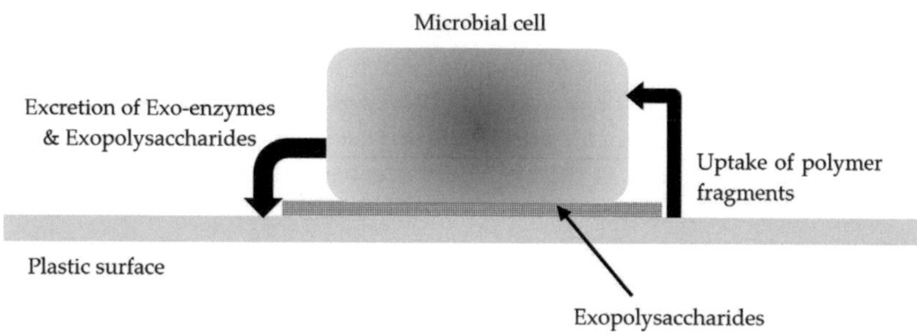

Figure 3.
Microbial attachment processes to a polymer surface [32].

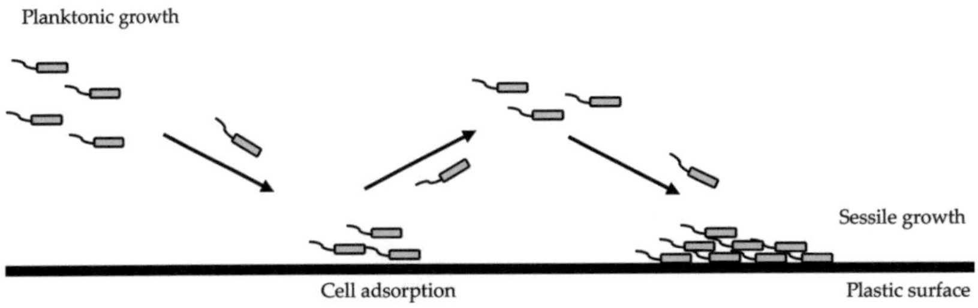

Figure 4.
Biofilm formation and processes [34].

environments, a broad spectrum of enzymatic activity is generally possible with wide functionalities. Biofilm formation can be assisted by the presence of pollutant chemical available at the polymer surface. The converse is also possible where surfaces contaminated with certain chemicals can prohibit biofilm formation. Biofilms continue to grow with the input of fresh nutrients, but when nutrients are deprived, the films will detach from the surface and return to a planktonic mode of growth. Overall hydrophobicity of the polymer surface and the surface charge of a bacterium may provide a reasonable prediction of surfaces to which a microorganism might colonize [37]. These initial cell-surface and cell-cell interactions are very useful to biofilm formation but incomplete (**Figure 4**). Microbial surfaces are heterogeneous, and can change widely in response to environmental changes. Five stages of biofilm development: have been identified as (1) initial attachment, (2) irreversible attachment, (3) maturation I, (4) maturation II, and (5) dispersion. Further research is required to provide the understanding of microbial components involved in biofilm development and regulation of their production to assemble to various facets of this complex microbial phenomenon [38].

The activities envisioned in this scenario (depicted in **Figure 4**) are the reversible adsorption of bacteria occurring at the later time scale, irreversible attachment of bacteria occurring at the second-minute time scale, growth and division of bacteria in hours-days, exopolymer production and biofilm formation in hours-days, and attachment and other organisms to biofilm in days-months.

3.5 Standardized testing methods

The evaluation of the extent of polymer biodegradation is made difficult by the dependence on polymer surface and the departure of degradation kinetics from the techniques available for small pollutant molecule techniques [39]. For applications for polymer biodegradation a variety of techniques have been applied. Visual observations, weight loss measurements, molar mass and mechanical properties, carbon dioxide evolution and/or oxygen consumption, radiolabeling, clear-zone formation, enzymatic degradation, and compost test under controlled conditions have been cited for their utility [27]. The testing regime must be explicitly described within a protocol of steps that can be collected for various polymers and compared on an equal basis. National and international efforts have developed such protocols to enable the desired comparisons using rigorous data collecting techniques and interpretation [40].

4. Environmental biodegradation of polymers

The conventional polymers such as (PE), (PP), (PS), (PUR), and (PET) are recognized for their persistence in the environment [41]. Each of these polymers is subject to very slow fragmentation to form small particles in a process expected to require centuries of exposure to photo-, physical, and biological degradation processes. Until recently, the commercial polymers were not expected to biodegrade. The current perspective supports polymer biodegradation with hopeful expectation that these newly encountered biodegradation processes can be transformed into technologies capable of providing major assistance to the ongoing task of waste polymer management.

4.1 Polyolefins

The polyolefins such as polyethylene (PE) have been recognized as a polymer remarkably resistant to degradation [42]. Products made with PE are very diverse and a testament to its chemical and biological inertness. The biodegradation of the polyolefins is complex and incompletely understood. Pure strains elicited from the environment have been used to investigate metabolic pathways or to gain a better understanding of the effect that environmental conditions have on polyolefin degradation. This strategy ignores the importance of different microbial species that could participate in a cooperative process. Treatment of the complex environments associated with polymeric solid waste could be difficult with information based on pure strain analysis. Mixed and complex microbial communities have been used and encountered in different bioremediation environments [43].

A variety of common PE types, low-density PE (LDPE), high-density PE (HDPE), linear low-density PE (LLDPE) and cross-linked PE (XLPE), differ in their density, degree of branching and availability of functional groups at the surface. The type of polymer used as the substrate can strongly influence the microbial community structure colonizing PE surface. A significant number of microbial strains have been identified for the deterioration caused by their interaction with the polymer surface [44]. Microorganisms have been categorized for their involvement in PE colonization and biodegradation or the combination. Some research studies did not conduct all the tests required to verify PE biodegradation. A more inclusive approach to assessing community composition, including the non-culturable fraction of microorganisms invisible by traditional microbiology methods is required in future assessments. The diversity of microorganisms capable of degrading PE extends beyond 17 genera of bacteria and nine genera of fungi [45]. These numbers are expected to increase with the use of more sensitive isolation and characterization techniques using rDNA sequencing. Polymer additives can affect the kinds of microorganisms colonizing the surfaces of these polymers. The ability of microorganisms to colonize the PE surfaces exhibits a variety of effects on polymer properties. Seven different characteristics have been identified and are used to monitor the extent of polymer surface change resulting from biodegradation of the polymer. The characteristics are hydrophobicity/hydrophilicity, crystallinity, surface topography, functional groups on the surface, mechanical properties, and molecular weight distribution. The use of surfactants has become important to PE biodegradation. Complete solubilization of PE in water by a *Pseudomonas fluorescens* treated for a month followed by biosurfactant treatment for a subsequent month in the second month and finally a 10% sodium dodecyl sulfate treatment at 60°C for a third month led to complete polymer degradation. A combination of *P. fluorescens*, surfactant and biosurfactant treatments as a single treatment significantly exhibited polymer oxidation and biodegradation [46]. The metabolically diverse genus *Pseudomonas* has been investigated for its capabilities to degrade and metabolize synthetic plastics. *Pseudomonas* species found in environmental matrices have been identified to degrade a variety of polymers including PE, and PP [47]. The unique capabilities of *Pseudomonas* species related to degradation and metabolism of synthetic polymers requires a focus on: the interactions controlling cell surface attachment of biofilms to polymer surfaces, extracellular polymer oxidation and/or hydrolytic enzyme activity, metabolic pathways mediating polymer uptake and degradation of polymer fragments within the microbial cell through catabolism, and the importance of development of the implementation of enhancing factors such as pretreatments, microbial consortia and nutrient availability while minimizing the effects of constraining factors such as alternative carbon sources and inhibitory by-products. In an ancillary study, thermophilic consortia of *Brevibacillus* sps. and *Aneurinibacillus* sp. from waste management landfills and sewage treatment plants exhibited enhanced PE and PP degradation [48].

The larval stage of two waxworm species, *Galleria mellonella* and *Plodia interpunctella*, has been observed to degrade LDPE without pretreatment [49, 50]. The worms could macerate PE as thin film shopping bags and metabolize the film to ethylene glycol which in turn biodegrades rapidly. The remarkable ability to digest a polymer considered non-edible may parallel the worm's ability utilize beeswax as a food source. From the guts of *Plodia interpunctella* waxworms two strains of bacteria, *Enterobacter asburiae* YP1 and *Bacillus* sp. YP1, were isolated and found to degrade PE in laboratory conditions. The two strains of bacteria were shown to reduce the polymer film hydrophobicity during a 28-day incubation. Changes to the film surface as cavities and pits were observed using scanning electron microscopy and atomic-force microscopy. Simple contact of ~100 *Galleria mellonella* worms with a commercial PE shopping bag for 12 hours resulted in a mass loss of 92 mg. The waxworm research has been scrutinized and found to be lacking the necessary information to support the claims of the original *Galleria mellonella* report [51].

Polypropylene (PP) is very similar to PE, in solution behavior and electrical properties. Mechanical properties and thermal resistance are improved with the addition of the methyl group but chemical resistance decreases. There are three forms of propylene selectively formed from the monomer isotactic, syndiotactic, and atactic due to the different geometric relationships achievable through polymerization technology. PP properties are strongly directed by tacticity or the methyl group orientation as related the methyl groups in neighboring monomer units. Isotactic PP has a greater degree of crystallinity than atactic and syndiotactic PP and therefore more difficult to biodegrade. The high molar mass of PP prohibits permeation through the microbial cell membrane which thwarts metabolism by living organisms. It is generally recognized that abiotic degradation provides a foothold for microorganisms to form a biofilm. With partial destruction of the polymer surface by abiotic effects the microbes can then start breaking the damaged polymer chains [52].

4.2 Polystyrene

PS is a sturdy thermoplastic commonly used in short-lifetime items that contribute broadly to the mass of poorly controlled polymers [53]. Various forms of PS such as general purpose (GPPS)/oriented polystyrene (OPS), polystyrene foam, and expanded polystyrene (EPS) foam are available for different commercial leading to a broad solid waste composition. PS has been thought to be non-biodegradable. The rate of biodegradation encountered in the environment is very slow leading to prolonged persistence as solid waste. In the past, PS was recycled through mechanical, chemical, and thermal technologies yielding gaseous and liquid daughter products [54]. A rather large collection of studies has shown that PS is subject to biodegradation but at a very slow rate in the environment. A sheet of PS buried for 32 years. in soil showed no indication of biotic or abiotic degradation [55]. The hydrophobicity of the polymer surface, a function of molecular structure and composition, detracts from the effectiveness of microbial attachment [56, 57]. The general lack of water solubility of PS prohibits the transport into microbial cells for metabolism.

A narrow range of microorganisms have been elicited for the environment and found to degrade PS [53]. *Bacillus* and *Pseudomonas* strains isolated from soil samples have been shown to degrade brominated high impact PS. The activity was seen in weight loss and surface changes to the PS film. Soil invertebrates such as the larvae of the mealworm (*Tenebrio molitor* Linnaeus) have been shown to chew and eat Styrofoam [57]. Samples of the larvae were fed Styrofoam as the sole diet for 30 days and compared with worms fed a conventional diet. The worms feeding Styrofoam survived for 1 month after which they stopped eating as they entered the pupae stage and emerged as adults after a subsequent 2 weeks. It appears that Styrofoam feeding did not lead to any lethality for the mealworms. The ingested PS mass was efficiently depolymerized within the larval gut during the retention time of 24 hours and converted to CO_2 [51]. This remarkable behavior by the mealworm can be considered the action of an efficient bioreactor. The mealworm can provide all the necessary components for PS treatment starting with chewing, ingesting, mixing, reacting with gut contents, and microbial degradation by gut microbial consortia. A PS-degrading bacterial strain *Exiguobacterium* sp. strain YT2 was isolated from the gut of mealworms and found to degrade PS films outside the mealworm gut. Superworms (*Zophobas morio*) were found to exhibit similar activity toward Styrofoam. Brominated high impact polystyrene (blend of polystyrene and polybutadiene) has been found to be degraded by *Pseudomonas* and *Bacillus* strains [58]. In a complementary study, four non-pathogenic cultures (*Enterobacter* sp., *Citrobacter sedlakii*, *Alcaligenes* sp. and *Brevundimonas diminuta*) were isolated from partially degraded polymer samples from a rural market setting and each were found to degrade high impact polystyrene [59].

4.3 Polyvinyl chloride

PVC is manufactured in two forms rigid and flexible. The rigid form can be found in the construction industry as pipe or in structural applications. The soft and flexible form can be made through the incorporation of plasticizers such as phthalates. Credit cards, bottles, and non-food packaging are notable products with a PVC composition. PVC has been known from its inception as a polymer with remarkable resistance to degradation [60]. Thermal and photodegradation processes are widely recognized for their role in the weathering processes found with PVC [61, 62]. The recalcitrant feature of polyvinyl chloride resistance to biodegradation becomes a matter of environmental concern across the all processes extending from manufacturing to waste disposal. Few reports are available relating the extent of PVC biodegradation. Early studies investigated the biodegradation of low-molecular weight PVC by white rot fungi [63]. Plasticized PVC was found to be degraded by fungi such as *As. fumigatus*, *Phanerochaete chrysosporium*, *Lentinus tigrinus*, *As. niger*, and *Aspergillus sydowii* [64].

Modifying the PVC film composition with adjuvants such as cellulose and starch provided a substrate that fungi could also degrade [65]. Several investigations of soil bacteria for the ability to degrade PVC from enrichment cultures were conducted on different locations [66]. Mixed cultures containing bacteria and fungi were isolated and found to grow on plasticized PVC [67]. Significant differences were observed for the colonization by the various components of the mixed isolates during very long exposure times [68]. Significant drift in isolate activity was averted through the use of talc. Consortia composed of a combination of different bacterial strains of *Pseudomonas otitidis*, *Bacillus cereus*, and *Acanthopleurobacter pedis* have the ability to degrade PVC in the environment [64]. These results offer the opportunity to optimization conditions for consortia growth in PVC and use as a treatment technology to degrade large collections of PVC. PVC film blends were shown to degrade by partnering biodegradable polymers with PVC [69].

4.4 Polyurethane

PUR encompass a broad field of polymer synthesis where a di- or polyisocyanate is chemically linked through carbamate (urethane) formation. These thermosetting and thermoplastic polymers have been utilized to form microcellular foams, high performance adhesives, synthetic fibers, surface coatings, and automobile parts along with a myriad of other applications. The carbamate linkage can be severed by chemical and biological processes [70].

Aromatic esters and the extent of the crystalline fraction of the polymer have been identified as important factors affecting the biodegradation of PUR [71, 72]. Acid and base hydrolysis strategies can sever the carbamate bond of the polymer. Microbial ureases, esterases and proteases can enable the hydrolysis the carbamate and ester bonds of a PUR polymer [71, 73, 74]. Bacteria have been found to be good sources for enzymes capable of degrading PUR polymers [75–82]. Fungi are also quite capable of degrading PUR polymers [83–85]. Each of the enzyme systems has their preferential targets: ureases attack the urea linkages [86–88] with esterases and proteases hydrolyzing the ester bonds of the polyester PUR as a major mechanism for its enzymatic depolymerization [89–92]. PUR polymers appear to be more amenable to enzymatic depolymerization or degradation but further searches and inquiry into hitherto unrecognized microbial PUR degrading activities is expected to offer significant PUR degrading activities.

4.5 Polyethylene terephthalate

PET is a polyester commonly marketed as a thermoplastic polymer resin finding use as synthetic fibers in clothing and carpeting, food and liquid containers, manufactured objects made through thermoforming, and engineering resins with glass fiber. Composed of terephthalic acid and ethylene glycol through the formation of ester bonds, PET has found a substantial role in packaging materials, beverage bottles and the textile industry. Characterized as a recalcitrant polymer of remarkable durability, the polymer's properties are reflective of its aromatic units in its backbone and a limited polymer chain mobility [91]. In many of its commercial forms, PET is semicrystalline having crystalline and amorphous phases which has a major effect on PET biodegradability. The environmental accumulation of PET is a testament of its versatility and the apparent lack of chemical/physical mechanisms capable of attacking its structural integrity show it to be a major environmental pollution problem.

The durability and the resulting low biodegradability of PET are due to the presence of repeating aromatic terephthalate units in its backbone and the corresponding limited mobility of the polymer chains [92]. The semicrystalline PET polymer also contains both amorphous and crystalline fractions with a strong effect on its biodegradability. Crystallinity exceeding 30% in PET beverage bottles and fibers having even higher crystalline compositions presents major hurdles to enzyme-induced degradation [93, 94]. At higher temperatures, the amorphous fraction of PET becomes more flexible and available to enzymatic degradation [95, 96]. The hydrolysis of PET by enzymes has been identified as a surface erosion process [97–100]. The hydrophobic surface significantly limits biodegradation due to the limited ability for microbial attachment. The hydrophobic nature of PET poses a significant barrier to microbial colonization of the polymer surface thus attenuating effective adsorption and access by hydrolytic enzymes to accomplish the polymer degradation [101].

A wide array of hydrolytic enzymes including hydrolases, lipases, esterases, and cutinases has been shown to have the ability to hydrolyze amorphous PET polymers and modify PET film surfaces. Microbes from a vast collection of waste sites and dumping situations have been studied for their ability to degrade PET. A subunit of PET, diethylene glycol phthalate has been found to be a source of carbon and energy necessary to the sustenance of microbial life. Enzyme modification may be effectively employed to improve the efficiency and specificity of the polyester degrading enzymes acknowledged to be active degraders of PET [102]. Significant efforts have been extended to developing an understanding of the enzymatic activity of high-performing candidate enzymes through selection processes, mechanistic probes, and enzyme engineering. In addition to hydrolytic enzymes already identified, enzymes found in thermophilic anaerobic sludge were found to degrade PET copolymers formed into beverage bottles [103].

Recently, the discovery of microbial activity capable of complete degradation of widely used beverage bottle plastic expands the range of technology options available for PET treatment. A microorganism isolated from the area adjacent to a plastic bottle-recycling facility was shown to aerobically degrade PET to small molecular daughter products and eventually to CO_2 and H_2O. This new research shows that a newly isolated microbial species, *Ideonella sakaiensis* 201-F6, degrades PET through hydrolytic transformations by the action of two enzymes, which are extracellular and intracellular hydrolases. A primary hydrolysis reaction intermediate, mono (hydroxy-2-ethyl) terephthalate is formed and can be subsequently degraded to ethylene glycol and terephthalic acid which can be utilized by the microorganism for growth [104–109].

This discovery could be a candidate as a single vessel system that could competently accomplish PET hydrolysis as an enzyme reactor. This may be the beginning of viable technology development applicable to the solution of the global plastic problem recognized for its terrestrial component as well as the water contamination problem found in the sea. These remarkable discoveries offer a new perspective on

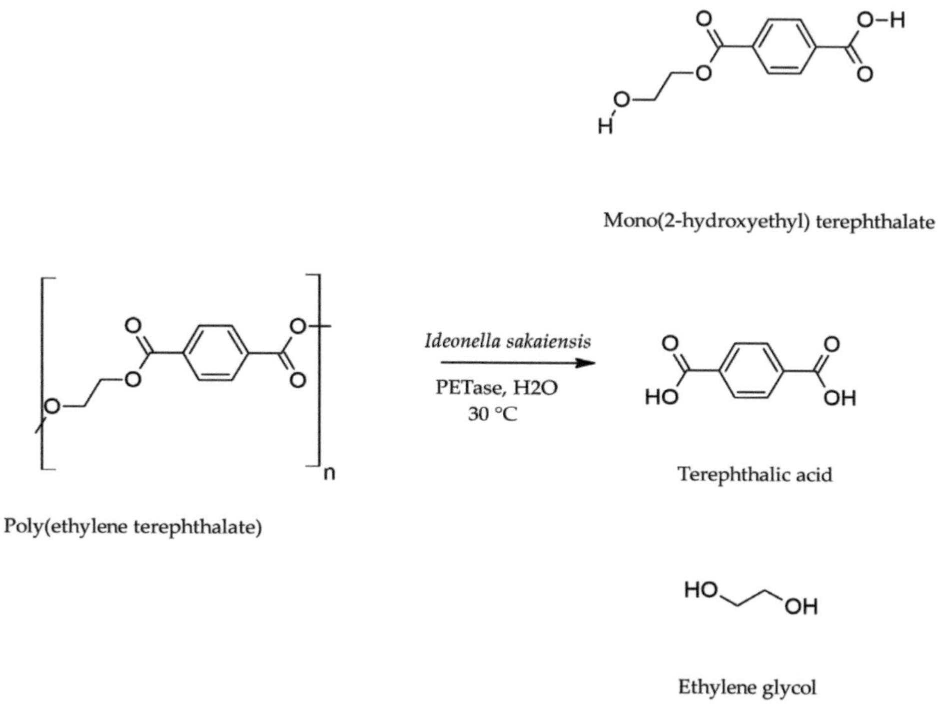

Figure 5.
Microbial depolymerization of poly(ethylene terephthalate).

the recalcitrant nature of PET and how future environmental management of PET waste may be conducted using the power of enzymes. The recognition of current limiting steps in the biological depolymerization of PET are expected to enable the design of a enzymes-based process to reutilized the natural assets contained in scrap PET [110] (**Figure 5**).

5. Conclusions

The major commercial polymers have been shown to be biodegradable in a variety of circumstances despite a strong predisposition suggesting that many of these polymers were recalcitrant to the effects of biodegradation. The question of whether bioremediation can play a significant role in the necessary management of polymer waste remains to be determined. Treatment technology for massive waste polymer treatment must be sufficiently robust to be reliable at large scale use and adaptable to conditions throughout the environment where this treatment is required. The status of information relating to the application of biodegradation treatment to existing and future polymer solid waste is at early stages of development for several waste polymers. The discovery of that invertebrate species (insect larvae) can reduce the size of the waste polymer by ingesting and degradation in the gut via enzymes which aid or complete degradation is rather amazing and requires additional scrutiny. There is an outside change that a polymer recycling technology based on these findings is a future possibility.

Disclaimer

The views expressed in this book chapter are those of the author and do not necessarily represent the views or policies of the U.S. Environmental Protection Agency.

Conflict of interest

No "conflict of interest" is known or expected.

Author details

John A. Glaser
U.S. Environmental Protection Agency, Office of Research and Development, National Risk Management Research Laboratory, Cincinnati, Ohio, USA

*Address all correspondence to: glaser.john@epa.gov

References

[1] Massy J. A Little Book about BIG Chemistry: The Story of Man-Made Polymers. New York: Springer; 2017. DOI: 10.1007/978-3-319-54831-9

[2] Wall LA, Flynn JH. Degradation of polymers. Rubber Chemistry and Technology. 1962;**35**(5):1157-1221. DOI: 10.5254/1.3539991

[3] David C. Thermal degradation of polymers. In: Comprehensive Chemical Kinetics. Vol. 14. Amsterdam: Elsevier; 1975. pp. 1-173. DOI: 10.1016/S0069-8040(08)70333-9

[4] Hawkins WL. Polymer degradation. In: Polymer Degradation and Stabilization. New York: Springer; 1984. pp. 3-34. DOI: 10.1007/978-3-642-69376-2_2

[5] Yousif E, Haddad R. Photodegradation and photostabilization of polymers, especially polystyrene: Review. Springer Plus. 2013;**2**:398. DOI: 10.1186/2193-1801-2-398

[6] Tokiwa Y, Calabia BP, Ugwu CU, Aiba S. Biodegradability of plastics. International Journal of Molecular Sciences. 2009;**10**(9):3722-3742. DOI: 10.3390/ijms10093722

[7] Ojeda T. Polymers and the environment. In: Yılmaz F, editor. Polymer Science. London: Intech Open; 2013. pp. 1-34. DOI: 10.5772/51057

[8] Geyer R, Jambeck JR, Law KL. Production, use, and fate of all plastics ever made. Science Advances. 2017;**3**(7). DOI: 10.1126/sciadv.1700782

[9] Hopewell J, Dvorak R, Kosior E. Plastics recycling: Challenges and opportunities. Philosophical Transactions of the Royal Society, B: Biological Sciences. 2009;**364**(1526):2115-2126

[10] Sperling LH. Introduction to Physical Polymer Science. New York: John Wiley & Sons; 2005. ISBN: 978-0-471-70606-9

[11] Namazi H. Polymers in our daily life. BioImpacts: BI. 2017;7(2):73. DOI: 10.15171/bi.2017.09

[12] Elias H-G. Macromolecules, Applications of Polymers. Vol. 4. New York: Wiley; 2009. ISBN: 978-3-527-31175-0

[13] Wypych G. Handbook of Material Weathering. Amsterdam: Elsevier; 2018. ISBN-13: 978-1927885314

[14] Brinson HF, Brinson LC. Polymer Engineering Science and Viscoelasticity. An Introduction. New York: Springer; 2008. DOI: 10.1007/978-1-4899-7485-3

[15] Van Krevelen DW, Te Nijenhuis K. Properties of Polymers: Their Correlation with Chemical Structure; Their Numerical Estimation and Prediction from Additive Group Contributions. Amsterdam: Elsevier; 2009. ISBN-13: 978-0080548197

[16] Zheng Y, Yanful EK, Bassi AS. A review of plastic waste biodegradation. Critical Reviews in Biotechnology. 2005;**25**(4):243-250. DOI: 10.1080/07388550500346359

[17] Wierckx N, Narancic T, Eberlein C, Wei R, Drzyzga O, Magnin A, et al. Plastic biodegradation: Challenges and opportunities. In: Consequences of Microbial Interactions with Hydrocarbons, Oils, and Lipids: Biodegradation and Bioremediation. New York: Springer; 2018. pp. 1-29. DOI: 10.1007/978-3-319-44535-9_23-1

[18] Banerjee A, Chatterjee K, Madras G. Enzymatic degradation of polymers: A brief review. Materials Science and Technology.

2014;**30**(5):567-573. DOI: 10.1179/1743284713Y.0000000503

[19] Kale SK, Deshmukh AG, Dudhare MS, Patil VB. Microbial degradation of plastic: A review. Journal of Biochemical Technology. 2015;**6**(2):952-961. ISSN: 0974-2328

[20] Pathak VM. Review on the current status of polymer degradation: A microbial approach. Bioresour Bioprocess. 2017;**4**(1):15. DOI: 10.1186/s40643-017-0145-9

[21] Ahmed T, Shahid M, Azeem F, Rasul I, Shah AA, Noman M, et al. Biodegradation of plastics: Current scenario and future prospects for environmental safety. Environmental Science and Pollution Research. 2018;**8**:7287-7298. DOI: 10.1007/s11356-018-1234-9

[22] Lucas N, Bienaime C, Belloy C, Queneudec M, Silvestre F, Nava-Saucedo JE. Polymer biodegradation: Mechanisms and estimation techniques—A review. Chemosphere. 2008;**73**(4):429-442. DOI: 10.1016/j.chemosphere.2008.06.064

[23] Gottschalk G. Bacterial Metabolism. New York: Springer Science & Business Media; 2012. ISBN-13: 978-0387961538

[24] Panikov NS. Microbial Growth Kinetics. New York: Springer Science & Business Media; 1995. ISBN-13: 978-0387913728

[25] Gumargalieva KZ, Zaikov GE. Biodegradation and Biodeterioration of Polymers: Kinetical Aspects. Hauppauge, NY: Nova Science Publishers; 1998. ISBN-13: 978-1560725824

[26] Mark HF. Encyclopedia of Polymer Science and Technology. 15 Volume Set. New York: Wiley; 2014. ISBN: 978-1-118-63389-2

[27] Shah AA, Hasan F, Hameed A, Ahmed S. Biological degradation of plastics: A comprehensive review. Biotechnology Advances. 2008;**26**(3):246-265. DOI: 10.1016/j.biotechadv.2007.12.005

[28] Guo W, Tao J, Yang C, Song C, Geng W, Li Q, et al. Introduction of environmentally degradable parameters to evaluate the biodegradability of biodegradable polymers. PLoS One. 2012;**7**(5):e38341. DOI: 10.1371/journal.pone.0038341

[29] Singh B, Sharma N. Mechanistic implications of plastic degradation. Polymer Degradation and Stability. 2008;**93**(3):561-584. DOI: 10.1016/j.polymdegradstab. 2007.11.008

[30] Azevedo HS, Reis RL. Understanding the enzymatic degradation of biodegradable polymers and strategies to control their degradation rate. In: Biodegradable Systems in Tissue Engineering and Regenerative Medicine. Boca Raton, FL: CRC Press; 2005. pp. 177-201. ISBN: 9780849319365

[31] O'Toole G, Kaplan HB, Kolter R. Biofilm formation as microbial development. Annual Review of Microbiology. 2000;**54**:49-79. DOI: 10.1146/annurev.micro.51.1.49

[32] Nwodo UU, Green E, Okoh AI. Bacterial exopolysaccharides: Functionality and prospects. International Journal of Molecular Sciences. 2012;**13**:14002-14015. DOI: 10.3390/ijms131114002

[33] Flemming HC, Wingender J. The biofilm matrix. Nature Reviews Microbiology. 2010;**8**(9):623-633. DOI: 10.1038/nrmicro2415

[34] Flemming HC, Wingender J, Szewzyk U, Steinberg P, Rice SA, Kjelleberg S. Biofilms: An emergent form of bacterial life. Nature Reviews Microbiology. 2016;**14**(9):563-575. DOI: 10.1038/nrmicro.2016.94

[35] Gu JD. Microbial colonization of polymeric materials for space applications and mechanisms of biodeterioration: A review. International Biodeterioration and Biodegradation. 2007;**59**(3):170-179. DOI: 10.1016/j.ibiod.2006.08.010

[36] Monds RD, O'Toole GA. The developmental model of microbial biofilms: Ten years of a paradigm up for review. Trends in Microbiology. 2009;**17**(2):73-87. DOI: 10.1016/j.tim.2008.11.001

[37] Flemming HC. Relevance of biofilms for the biodeterioration of surfaces of polymeric materials. Polymer Degradation and Stability. 1998;**59**(1-3):309-315. DOI: 10.1016/S0141-3910(97)00189-4

[38] Elias S, Banin E. Multi-species biofilms: Living with friendly neighbors. FEMS Microbiology Reviews. 2012;**36**(5):990-1004. DOI: 10.1111/j.1574-6976.2012.00325.x

[39] Funabashi M, Ninomiya F, Kunioka M. Biodegradability evaluation of polymers by ISO 14855-2. International Journal of Molecular Sciences. 2009;**10**(8):3635-3654

[40] Mueller RJ. Biodegradability of polymers: Regulations and methods for testing. In: Steinbüchel A, editor. Biopolymers, General Aspects and Special Applications. Vol. 10. Weinheim: Wiley-VCH; 2003. pp. 366-388. DOI: 10.1002/3527600035.bpola012

[41] Fotopoulou KN, Karapanagioti HK. Degradation of various plastics in the environment. In: The Handbook of Environmental Chemistry. New York: Springer; 2017. DOI: 10.1007/698_2017_11

[42] Agboola O, Sadiku R, Mokrani T, Amer I, Imoru O. Polyolefins and the environment. In: Polyolefin Fibres. 2nd ed. Amsterdam: Elsevier; 2017. pp. 89-133. DOI: 10.1016/B978-0-08-101132-4.00004-7

[43] Roy PK, Hakkarainen M, Varma IK, Albertsson AC. Degradable polyethylene: Fantasy or reality. Environmental Science & Technology. 2011;**45**(10):4217-4227. DOI: 10.1021/es104042f

[44] Restrepo-Flórez JM, Bassi A, Thompson MR. Microbial degradation and deterioration of polyethylene—A review. International Biodeterioration and Biodegradation. 2014;**88**:83-90. DOI: 10.1016/j.ibiod.2013.12.014

[45] Arkatkar A, Arutchelvi J, Sudhakar M, Bhaduri S, Uppara PV, Doble M. Approaches to enhance the biodegradation of polyolefins. The Open Environmental Engineering Journal. 2009;**2**:68-80. DOI: 10.2174/1874829500902010068

[46] Arkatkar A, Juwarkar AA, Bhaduri S, Uppara PV, Doble M. Growth of *Pseudomonas* and *Bacillus* biofilms on pretreated polypropylene surface. International Biodeterioration and Biodegradation. 2010;**64**(6):530-536. DOI: 10.1016/j.ibiod.2010.06.002

[47] Wilkes RA, Aristilde L. Degradation and metabolism of synthetic plastics and associated products by *Pseudomonas* spp.: Capabilities and challenges. Journal of Applied Microbiology. 2017;**123**:582-593. DOI: 10.1111/jam.13472

[48] Skariyachan S, Patil AA, Shankar A, Manjunath M, Bachappanavar N, Kiran S. Enhanced polymer degradation of polyethylene and polypropylene by novel thermophilic consortia of *Brevibacillus* sps. and *Aneurinibacillus* sp. screened from waste management landfills and sewage treatment plants. Polymer Degradation and Stability. 2018;**149**:52-68. DOI: 10.1016/j.polymdegradstab.2018.01.018

[49] Bombelli P, Howe CJ, Bertocchini F. Polyethylene bio-degradation by caterpillars of the wax moth *Galleria mellonella*. Current Biology. 2017;**27**(8):R292-R293. DOI: 10.1016/j.cub.2017.02.060

[50] Lwanga EH, Thapa B, Yang X, Gertsen H, Salánki T, Geissen V, et al. Decay of low-density polyethylene by bacteria extracted from earthworm's guts: A potential for soil restoration. The Science of the Total Environment. 2018;**624**:753-757. DOI: 10.1016/j.scitotenv.2017.12.144

[51] Yang Y, Yang J, Wu WM, Zhao J, Song Y, Gao L, et al. Biodegradation and mineralization of polystyrene by plastic-eating mealworms: Part 2. Role of gut microorganisms. Environmental Science & Technology. 2015;**49**(20):12087-12093

[52] Celso Luis de Carvalho CL, Derval dos Santos Rosa D. Chapter 5: Polypropylene biodegradation. In: Garcia P, editor. Polypropylene—Properties, Uses and Benefits. New York: Nova Science Publishers, Inc; 2016. pp. 141-174. ISBN-13: 978-1536100082

[53] Ho BT, Roberts TK, Lucas S. An overview on biodegradation of polystyrene and modified polystyrene: The microbial approach. Critical Reviews in Biotechnology. 2018;**38**(2):308-320. DOI: 10.1080/07388551.2017.1355293

[54] Maharana T, Negi YS, Mohanty B. Recycling of polystyrene. Polymer-Plastics Technology and Engineering. 2007;**46**(7):729-736. DOI: 10.1080/03602550701273963

[55] Otake Y, Kobayashi T, Asabe H, Murakami N, Ono K. Biodegradation of low-density polyethylene, polystyrene, polyvinyl chloride, and urea formaldehyde resin buried under soil for over 32 years. Journal of Applied Polymer Science. 1995;**56**(13):1789-1796. DOI: 10.1002/app.1995.070561309

[56] Kowalczyk A, Chyc M, Ryszka P, Latowski D. *Achromobacter xylosoxidans* as a new microorganism strain colonizing high-density polyethylene as a key step to its biodegradation. Environmental Science and Pollution Research. 2016;**23**(11):11349-11356. DOI: 10.1007/s11356-016-6563-y

[57] Yang Y, Yang J, Wu WM, Zhao J, Song Y, Gao L, et al. Biodegradation and mineralization of polystyrene by plastic-eating mealworms: Part 1. Chemical and physical characterization and isotopic tests. Environmental Science & Technology. 2015;**49**(20):12080-12086

[58] Sekhar VC, Nampoothiri KM, Mohan AJ, Nair NR, Bhaskar T, Pandey A. Microbial degradation of high impact polystyrene (H;IPS), an e-plastic with decabromodiphenyl oxide and antimony trioxide. Journal of Hazardous Materials. 2016;**318**:347-354. DOI: 10.1016/j.jhazmat.2016.07.008

[59] Braun D. Poly (vinyl chloride) on the way from the 19th century to the 21st century. Journal of Polymer Science, Part A: Polymer Chemistry. 2004;**42**(3):578-586. DOI: 10.1002/pola.10906

[60] Owen J. Degradation and Stabilisation of PVC. New York: Springer Science & Business Media; 2012. ISBN-13: 978-9401089784

[61] Allsopp MW, Vianello G. Vinyl chloride polymers. In: Encyclopedia of Polymer Science and Technology. Wiley; 2003. DOI: 10.1002/0471440264.pst386

[62] Kırbaş Z, Keskin N, Güner A. Biodegradation of polyvinylchloride (PVC) by white rot fungi. Bulletin of Environmental Contamination and Toxicology. 1999;**63**(3):335-342. DOI: 10.1007/s001289900985

[63] Ali MI, Ahmed S, Robson G, Javed I, Ali N, Atiq N, et al. Isolation and

molecular characterization of polyvinyl chloride (PVC) plastic degrading fungal isolates. Journal of Basic Microbiology. 2014;**54**(1):18-27. DOI: 10.1002/jobm.201200496

[64] Ali MI, Ahmed S, Javed I, Ali N, Atiq N, Hameed A, et al. Biodegradation of starch blended polyvinyl chloride films by isolated *Phanerochaete chrysosporium* PV1. International Journal of Environmental Science and Technology. 2014;**11**(2):339-348. DOI: 10.1007/s13762-013-0220-5

[65] Anwar MS, Kapri A, Chaudhry V, Mishra A, Ansari MW, Souche Y, et al. Response of indigenously developed bacterial consortia in progressive degradation of polyvinyl chloride. Protoplasma. 2016;**253**(4):1023-1032. DOI: 10.1007/s00709-015-0855-9

[66] Patil R, Bagde US. Isolation of polyvinyl chloride degrading bacterial strains from environmental samples using enrichment culture technique. African Journal of Biotechnology. 2012;**11**(31):7947-7956. DOI: 10.5897/AJB11.3630

[67] Pečiulytė D. Microbial colonization and biodeterioration of plasticized polyvinyl chloride plastics. Ekologija. 2002;**4**:7-15. ISSN: 0235-7224

[68] Sah A, Negi H, Kapri A, Anwar S, Goel R. Comparative shelf life and efficacy of LDPE and PVC degrading bacterial consortia under bioformulation. Ekologija. 2011;**57**(2). DOI: 10.6001/ekologija.v57i2.1885

[69] Sonnenschein MF. Polyurethanes: Science, Technology, Markets, and Trends. New York: John Wiley & Sons; 2014. ISBN-13: 978-1118737835

[70] Urgun-Demirtas M, Singh D, Pagilla K. Laboratory investigation of biodegradability of a polyurethane foam under anaerobic conditions. Polymer Degradation and Stability. 2007;**92**:1599-1610. DOI: 10.1016/j.polymdegradstab. 2007.04.013

[71] Cregut M, Bedas M, Durand MJ, Thouand G. New insights into polyurethane biodegradation and realistic prospects for the development of a sustainable waste recycling process. Biotechnology Advances. 2013;**31**(8):1634-1647. DOI: 10.1016/j.biotechadv.2013.08.011

[72] Howard GT. Polyurethane biodegradation. In: Microbial Degradation of Xenobiotics. New York: Springer; 2012. pp. 371-394. DOI: 10.1007/978-3-642-23789-8_14

[73] Loredo-Treviño A, Gutiérrez-Sánchez G, Rodríguez-Herrera R, Aguilar CN. Microbial enzymes involved in polyurethane biodegradation: A review. Journal of Polymers and the Environment. 2012;**20**(1):258-265. DOI: 10.1007/s10924-011-0390-5

[74] Nakajima-Kambe T, Onuma F, Kimpara N, Nakahara T. Isolation and characterization of a bacterium which utilizes polyester polyurethane as a sole carbon and nitrogen source. FEMS Microbiology Letters. 1995;**129**(1):39-42. DOI: 10.1111/j.1574-6968.1995.tb07554.x

[75] Howard GT, Blake RC. Growth of *Pseudomonas fluorescens* on a polyester–polyurethane and the purification and characterization of a polyurethanase–protease enzyme. International Biodeterioration and Biodegradation. 1998;**42**(4):213-220. DOI: 10.1016/S0964-8305(98)00051-1

[76] Stern RV, Howard GT. The polyester polyurethanase gene (pueA) from *Pseudomonas chlororaphis* encodes a lipase. FEMS Microbiology Letters. 2000;**185**(2):163-168. DOI: 10.1111/j.1574-6968.2000.tb09056.x

[77] Howard GT, Crother B, Vicknair J. Cloning, nucleotide

sequencing and characterization of a polyurethanase gene (pueB) from *Pseudomonas chlororaphis*. International Biodeterioration and Biodegradation. 2001;**47**(3):141-149. DOI: 10.1016/S0964-8305(01)00042-7

[78] Howard GT. Biodegradation of polyurethane: A review. International Biodeterioration and Biodegradation. 2002;**49**(4):245-252. DOI: 10.1016/S0964-8305(02)00051-3

[79] Schmidt J, Wei R, Oeser T, Dedavid e Silva LA, Breite D, Schulze A, et al. Degradation of polyester polyurethane by bacterial polyester hydrolases. Polymers. 2017;**9**(2):65. DOI: 10.3390/polym9020065

[80] Shah AA, Hasan F, Akhter JI, Hameed A, Ahmed S. Degradation of polyurethane by novel bacterial consortium isolated from soil. Annales de Microbiologie. 2008;**58**(3):381. DOI: 10.1007/BF03175532

[81] Pathirana RA, Seal KJ. Studies on polyurethane deteriorating fungi. International Biodeterioration and Biodegradation. 1985;**21**:41-49

[82] Crabbe JR, Campbell JR, Thompson L, Walz SL, Schultz WW. Biodegradation of a colloidal ester-based polyurethane by soil fungi. International Biodeterioration and Biodegradation. 1994;**33**(2):103-113. DOI: 10.1016/0964-8305(94)90030-2

[83] Russell JR, Huang J, Anand P, Kucera K, Sandoval AG, Dantzler KW, et al. Biodegradation of polyester polyurethane by endophytic fungi. Applied and Environmental Microbiology. 2011;**77**(17):6076-6084. DOI: 10.1128/AEM.00521-11

[84] Owen S, Otani T, Masaoka S, Ohe T. The biodegradation of low-molecular-weight urethane compounds by a strain of *Exophiala jeanselmei*. Bioscience, Biotechnology, and Biochemistry. 1996;**60**(2):244-248. DOI: 10.1271/bbb.60.244

[85] Ruiz C, Main T, Hilliard NP, Howard GT. Purification and characterization of twopolyurethanase enzymes from *Pseudomonas chlororaphis*. International Biodeterioration and Biodegradation. 1999;**43**(1-2):43-47. DOI: 10.1016/S0964-8305(98)00067-5

[86] Matsumiya Y, Murata N, Tanabe E, Kubota K, Kubo M. Isolation and characterization of an ether-type polyurethane-degrading micro-organism and analysis of degradation mechanism by *Alternaria* sp. Journal of Applied Microbiology. 2010;**108**(6):1946-1953. DOI: 10.1111/j.1365-2672.2009.04600.x

[87] Nakajima-Kambe T, Shigeno-Akutsu Y, Nomura N, Onuma F, Nakahara T. Microbial degradation of polyurethane, polyester polyurethanes and polyether polyurethanes. Applied Microbiology and Biotechnology. 1999;**51**(2):134-140. DOI: 10.1007/s002530051

[88] Tang YW, Labow RS, Santerre JP. Enzyme-induced biodegradation of polycarbonate-polyurethanes: Dependence on hard-segment chemistry. Journal of Biomedical Materials Research Part A. 2001;**57**(4):597-611. DOI: 10.1002/1097-4636(20011215)57:4<597::AID-JBM1207>3.0.CO;2-T

[89] Tang YW, Labow RS, Santerre JP. Enzyme-induced biodegradation of polycarbonate polyurethanes: Dependence on hard-segment concentration. Journal of Biomedical Materials Research Part A. 2001;**56**(4):516-528. DOI: 10.1002/1097-4636(20011215)57:4<597::AID-JBM1207>3.0.CO;2-T

[90] Krasowska K, Heimowska A, Rutkowska M. Environmental degradability of polyurethanes. In:

Thermoplastic Elastomers-Synthesis and Applications. InTech; 2015. DOI: 10.5772/60925

[91] Marten E, Müller RJ, Deckwer WD. Studies on the enzymatic hydrolysis of polyesters I. Low molecular mass model esters and aliphatic polyesters. Polymer Degradation and Stability. 2003;**80**(3):485-501. DOI: 10.1016/S0141-3910(03)00032-6

[92] Marten E, Müller RJ, Deckwer WD. Studies on the enzymatic hydrolysis of polyesters. II. Aliphatic–aromatic copolyesters. Polymer Degradation and Stability. 2005;**88**(3):371-381. DOI: 10.1016/j.polymdegradstab. 2004.12.001

[93] Liu J, Xu G, Dong W, Xu N, Xin F, Ma J, et al. Biodegradation of diethyl terephthalate (DET) and polyethylene terephthalate (PET) by a novel identified degrader *Delftia* sp. WL-3 and its proposed metabolic pathway. Letters in Applied Microbiology. 2018;**67**:254-261. DOI: 10.1111/lam.13014

[94] Lee JH, Lim KS, Hahm WG, Kim SH. Properties of recycled and virgin poly (ethylene terephthalate) blend fibers. Journal of Applied Polymer Science. 2013;**128**(2):1250-1256. DOI: 10.1002/app.38502

[95] Parikh M, Gross R. The effect of crystalline morphology on enzymatic degradation kinetics. In: Ching C, Kaplan DL, Thomas EL, editors. Biodegradable Polymers and Packaging. Lancaster-Basel: Technomic; 1993. pp. 407-415. ISBN-13: 978-1566760089

[96] Ronkvist ÅM, Xie W, Lu W, Gross RA. Cutinase-catalyzed hydrolysis of poly (ethylene terephthalate). Macromolecules. 2009;**42**(14): 5128-5138. DOI: 10.1021/ma9005318

[97] Zhang J, Wang X, Gong J, Gu Z. A study on the biodegradability of polyethylene terephthalate fiber and diethylene glycol terephthalate. Journal of Applied Polymer Science. 2004;**93**(3):1089-1096. DOI: 10.1002/app.20556

[98] Mueller R-J. Aliphatic–aromatic polyesters. In: Bastioli C, editor. Handbook of Biodegradable Polymers. Shawbury, UK: Rapra Technology Ltd; 2005. ISBN: 1-85957-389-4303-38

[99] Mueller RJ. Biological degradation of synthetic polyesters—Enzymes as potential catalysts for polyester recycling. Process Biochemistry. 2006;**41**(10):2124-2128. DOI: 10.1016/j.procbio.2006.05.018

[100] Wei R, Oeser T, Barth M, Weigl N, Lübs A, Schulz-Siegmund M, et al. Turbidimetric analysis of the enzymatic hydrolysis of polyethylene terephthalate nanoparticles. Journal of Molecular Catalysis B: Enzymatic. 2014;**103**:72-78. DOI: 10.1016/j.molcatb.2013.08.010

[101] Atthoff B, Hilborn J. Protein adsorption onto polyester surfaces: Is there a need for surface activation? Journal of Biomedical Materials Research Part B, Applied Biomaterials. 2007;**80**(1):121-130. DOI: 10.1002/jbm.b.30576

[102] Biundo A, Ribitsch D, Guebitz GM. Surface engineering of polyester-degrading enzymes to improve efficiency and tune specificity. Applied Microbiology and Biotechnology. 2018;**102**(8):3551-3559. DOI: 10.1007/s0025

[103] Hermanová S, Šmejkalová P, Merna J, Zarevúcka M. Biodegradation of waste PET based copolyesters in thermophilic anaerobic sludge. Polymer Degradation and Stability. 2015;**111**:176-184. DOI: 10.1016/j.polymdegradstab. 2014.11.007

[104] Yoshida S, Hiraga K, Takehana T, Taniguchi I, Yamaji H, Maeda Y, et al. A bacterium that degrades and assimilates poly (ethylene terephthalate). Science. 2016;**351**(6278):1196-1199

[105] Yoshida S, Hiraga K, Takehana T, Taniguchi I, Yamaji H, Maeda Y, et al. Response to comment on "A bacterium that degrades and assimilates poly (ethylene terephthalate)". Science. 2016;**353**(6301):759

[106] Bornscheuer UT. Feeding on plastic. Science. 2016;**351**(6278): 1154-1155. DOI: 10.1126/science.aaf2853

[107] Joo S, Cho IJ, Seo H, Son HF, Sagong HY, Shin TJ, et al. Structural insight into molecular mechanism of poly (ethylene terephthalate) degradation. Nature Communications. 2018;**9**(1):382

[108] Doppalapudi S, Jain A, Khan W, Domb AJ. Biodegradable polymers—An overview. Polymers for Advanced Technologies. 2014;**25**(5):427-435. DOI: 10.1002/pat.3305

[109] Avérous L, Pollet E. Biodegradable polymers. In: Environmental Silicate Nano-Biocomposites. Springer; 2012. pp. 13-39. DOI: 10.1007/978-1-4471-4108-2_2

[110] Vroman I, Tighzert L. Biodegradable polymers. Materials. 2009;**2**(2):307-344. DOI: 10.3390/ma2020307